ITEM NO: 2025022

D1686140

A Practical Guide to Brain–Compu
with BCI2000

Gerwin Schalk · Jürgen Mellinger

A Practical Guide to Brain–Computer Interfacing with BCI2000

General-Purpose Software
for Brain–Computer Interface Research,
Data Acquisition, Stimulus Presentation,
and Brain Monitoring

 Springer

Gerwin Schalk
Wadsworth Center
New York State Department
  of Health
Albany, NY, USA
schalk@wadsworth.org

Jürgen Mellinger
Institute of Medical Psychology
  & Behavioral Neurobiology
University of Tübingen
Tübingen, Germany
juergen.mellinger@uni-tuebingen.de

ISBN 978-1-4471-5750-2   ISBN 978-1-84996-092-2 (eBook)
DOI 10.1007/978-1-84996-092-2
Springer London Dordrecht Heidelberg New York

British Library Cataloguing in Publication Data
A catalogue record for this book is available from the British Library

©Springer-Verlag London Limited 2010
Softcover re-print of the Hardcover 1st edition 2010
Apart from any fair dealing for the purposes of research or private study, or criticism or review, as permitted under the Copyright, Designs and Patents Act 1988, this publication may only be reproduced, stored or transmitted, in any form or by any means, with the prior permission in writing of the publishers, or in the case of reprographic reproduction in accordance with the terms of licenses issued by the Copyright Licensing Agency. Enquiries concerning reproduction outside those terms should be sent to the publishers.
The use of registered names, trademarks, etc., in this publication does not imply, even in the absence of a specific statement, that such names are exempt from the relevant laws and regulations and therefore free for general use.
The publisher makes no representation, express or implied, with regard to the accuracy of the information contained in this book and cannot accept any legal responsibility or liability for any errors or omissions that may be made.

Cover design: KünkelLopka GmbH, Heidelberg

Printed on acid-free paper

Springer is part of Springer Science+Business Media (www.springer.com)

The hope is that, in not too many years, human brains and computing machines will be coupled together very tightly, and that the resulting partnership will think as no human brain has ever thought and process data in a way not approached by the information-handling machines we know today.

J.C.R. Licklider
Man–Computer Symbiosis, March 1960

This book is dedicated to our wives Renate and Ulrike

Preface

What Is BCI2000?

BCI2000 is a general-purpose software platform for brain–computer interface (BCI) research. It can also be used for a wide variety of data acquisition, stimulus presentation, and brain monitoring applications. BCI2000 has been in development since 2000 in a project led by the Brain–Computer Interface R&D Program at the Wadsworth Center of the New York State Department of Health in Albany, New York, USA, with substantial contributions by the Institute of Medical Psychology and Behavioral Neurobiology at the University of Tübingen, Germany. In addition, many laboratories around the world, most notably the BrainLab at Georgia State University in Atlanta, Georgia, and Fondazione Santa Lucia in Rome, Italy, have also played an important role in the project's development.

Mission

The mission of the BCI2000 project is to facilitate research and the development of applications in all areas that depend on real-time acquisition, processing, and feedback of biosignals.

Vision

Our vision is that BCI2000 will become a widely used software tool for diverse areas of research and development.

History

In the late 1990s, Dr. Jonathan Wolpaw, a BCI pioneer at the Wadsworth Center, recognized the need for a software platform that would facilitate the implementation of any BCI design. This need was supported by Dr. Niels Birbaumer, another BCI pioneer from the University of Tübingen in Germany. Their joint interest resulted in a meeting that lasted for several days and took place in Tübingen in late 1999. It was attended by Gerwin Schalk and Dennis McFarland from the Wadsworth Center, and Thilo Hinterberger from the University of Tübingen. This meeting resulted in a first set of general system specifications. Because these system specifications were focused on technical flexibility but also practicality, they have not changed since then and there has not been a need to change them.

The next year and a half were characterized by the implementation of the first generation of the BCI2000 system, primarily by Gerwin Schalk and Dennis McFarland at the Wadsworth Center. These efforts culminated in the first successful BCI2000-based experiment in July 2001. The data files generated in that first set of experiments can still be interpreted today with the tools provided by BCI2000. Just like all other BCI2000 data files, these early data sets contained a full record of the parameterization of the experiments, and of the time course of important events such as the timing of stimulus presentation. Thus, it is possible to completely reconstruct all important details of experiments that were conducted many years ago.

During 2002, Jürgen Mellinger at the University of Tübingen joined the BCI2000 project and immediately began to improve the initial system implementation. The project also received initial exposure during the 2nd International BCI Meeting in Rensselaerville, NY, which was organized by the Wadsworth Center. This exposure resulted in the extension of the BCI2000 system to support a P300-based spelling device modeled after the methodology described in [7], and in the first adoption of the BCI2000 system by other research groups (e.g., Dr. Emanuel Donchin, University of South Florida; Dr. Scott Makeig, Swartz Center for Computational Neuroscience, La Jolla; Dr. Melody Moore-Jackson, Georgia State BrainLab). During 2004, Adam Wilson, then a graduate student at the University of Wisconsin at Madison and who has since worked as a postdoctoral associate at the Wadsworth Center, began to use BCI2000 and to contribute to the project.

Since the early years, the design and development of the BCI2000 platform have been subject to two contrasting forces that needed to be appropriately managed. These forces were the desire to develop a robust general-purpose BCI system (i.e., a long-term goal) on the one hand, and the necessity to support many different BCI experiments in Albany and Tübingen (i.e., short-term goals) on the other hand. In 2005, the continually improving BCI2000 platform and the initial success in the system's dissemination provided the basis for an NIH grant proposal dedicated to the further development and maintenance of the BCI2000 platform. (Initial BCI2000 development had been sponsored by an NIH Bioengineering Research Partnership (BRP) grant to Dr. Jonathan Wolpaw.) In 2006, this grant application was awarded for an initial four-year period by the National Institute of Biomedical Imaging and BioEngineering (NIBIB) of the NIH. This dedicated funding allowed the project team to focus entirely on the needs of the BCI research community rather than simply the needs of individual research projects. The resulting work culminated in the release of BCI2000 V2.0 in early 2008. This version represented the largest single advance since the project's inception. In V2.0, we consolidated and expanded the existing BCI2000 platform and associated development and maintenance processes to achieve a consistent set of quality core components, comprehensive and up-to-date documentation for users and software engineers, and testing and release management procedures. Since the release of version 2.0, the adoption of the BCI2000 platform has accelerated further. Many scientists and engineers are now requesting BCI2000 for purposes other than BCI research.

As we are writing this book, BCI2000 V3.0 is about to be released. This system increases performance (particularly on machines with multiple processor cores), supports VisualStudio and MinGW in addition to Borland/CodeGear compilers, and further modularizes BCI2000 components.

Impact to Date

BCI2000 has already had a substantial impact on BCI and related research. As of December 2009, BCI2000 has been acquired by nearly 500 laboratories around the world. The original article that described the BCI2000 system [19] has been cited more than 200 times, and was recently awarded a Best Paper Award of 2004 by *IEEE Transactions on Biomedical Engineering*. Furthermore, a review of the literature revealed that BCI2000 has been used in studies reported in more than 120 peer-reviewed publications. These publications include some of the most impressive BCI demonstrations and applications reported to date. E.g.: the first online brain–computer interfaces using magnetoencephalographic (MEG) signals [15] or electrocorticographic (ECoG) signals [8, 11, 12, 23]; the first multi-dimensional BCI using ECoG signals [22]; the first applications of BCI technology toward restoration of function in patients with chronic stroke [3, 24]; the use of BCI techniques to control assistive technologies [6]; the first real-time BCI use of high-resolution EEG techniques [5]; demonstrations that non-invasive BCI systems can support multi-dimensional cursor movements without [25, 26] and with [14] selection capabilities; control of a humanoid robot by a noninvasive BCI [2]; and the first demonstration that people severely paralyzed by amyotrophic lateral sclerosis (ALS) can operate a sensorimotor rhythm-based BCI [10]. BCI2000 is also supporting the only existing long-term in-home application of BCI technology for people who are severely disabled. In these ongoing studies by Jonathan Wolpaw and Theresa Vaughan at the Wadsworth Center, BCI2000-based systems are being placed in the homes of severely disabled people. For the past several years, these individuals have been using the BCI for word processing, email, environmental control, and daily communication with family and friends.

Many studies have used BCI2000 in fields related to BCI research. This includes the first large-scale motor mapping studies using ECoG signals [13, 17]; real-time mapping of cortical function using ECoG [16, 21]; the optimization of BCI signal processing routines [4, 18, 27]; evaluation of steady-state visual evoked potentials (SSVEP) for BCI purposes [1]; and the demonstration that two-dimensional hand movements and finger movements can be decoded from ECoG signals ([20] and [9], respectively). Facilitated by the easy exchange of data and experimental paradigms that BCI2000 enables, a number of these studies were performed as collaborations among several geographically widespread laboratories. To our knowledge, there have been no comparable large-scale collaborative BCI studies that have not used BCI2000.

Furthermore, BCI2000 has also been: used in demonstrations on national television including NBC, CBS, and CNN; referenced several hundred times in journal articles, media articles, and personal blogs; used or cited in dozens of Masters Theses or Doctoral Dissertations; listed as a significant qualification in curricula vitae; and even mentioned as desirable experience in job postings. The widespread and continually growing success of the BCI2000 platform is strong evidence for its utility.

In summary, BCI2000 is promoting and facilitating the evolution of BCI research and development from isolated laboratory demonstrations into clinically relevant BCI systems and applications useful to people with severe disabilities. As indicated by the above descriptions of its utility for many different aspects of BCI research, by its wide dissemination, and by its prominence in the scientific literature, BCI2000 is fast becoming, or perhaps has already become, the standard software platform for BCI research.

Dissemination

The BCI2000 software is available free of charge for research and educational purposes at http://www.bci2000.org. This web site contains comprehensive project-related information including additional documentation on a wiki and a bulletin board. In addition, the BCI2000 project has organized a number of workshops on the theory and application of the platform: Albany, New York, June 2005; Beijing, China, July 2007; Rome, Italy, December 2007; Utrecht, The Netherlands, July 2008; Bolton Landing, New York, October 2009; and Beijing, China, December 2009.

BCI2000 Benefits

Implementation of real-time software that integrates data acquisition, signal processing, and feedback is complex and difficult. BCI2000 is a platform in which the major technical difficulties have been solved. Thus, it allows a scientist or engineer to spend more time on their research and less time on validating and trouble-shooting the technology. In addition, BCI2000 offers several other important benefits:

- **An Established Solution** BCI2000 comes with proven support for different data acquisition hardware, signal processing routines, and experimental paradigms.
- **Facilitates Operation of Research Programs** Although a number of software platforms, such as Matlab or LabView, can be used to prototype experimental

paradigms, such prototypes do not have the common data format, software interfaces, or documenting protocols, that are important or even critical for success in
large research programs. In contrast, BCI2000 has been designed from the start
and developed over many years to support large research programs with many
diverse research projects.

- **Facilitates Deployment in Multiple Sites** The BCI2000 platform does not rely
  on 3rd-party software components for its operation. Even for compilation, it requires only affordable or free C++ compilers. Thus, both development and deployment of BCI2000 on multiple computers in multiple sites is very economical.
- **Cross-platform/compiler Compatibility** BCI2000 currently requires Microsoft
  Windows to operate and Borland's C++ Builder for compilation. BCI2000 V3.0
  also supports VisualStudio and MinGW.
- **Open License** BCI2000 is free and without restrictions for academic and research
  purposes.

Acknowledgments

Core Team

Project Leader	Gerwin Schalk, Ph.D.
Chief Software Engineer	Jürgen Mellinger, M.E.
Quality Control and Testing	Adam Wilson, Ph.D.

Additional Contributors and Acknowledgments

We would like to acknowledge the critical role of Dr. Dennis McFarland in the development of the initial system specifications and implementations, the support and
important advice of Drs. Wolpaw and Birbaumer in earlier stages of this project, the
continuing support by Theresa Vaughan and Dr. Jonathan Wolpaw throughout this
project, the support and advice by colleagues at Health Research and the Wadsworth
Center, in particular Bob Gallo and Erin Davis, Dr. Melody Moore-Jackson's help
with supervising of students who contributed to BCI2000, and Dr. Wilson's out-
standing efforts in commenting on and editing the book, which led to substantial
improvements. We are also extremely grateful to Peter Brunner for his sustained
and excellent technical assistance with many aspects relating to the BCI2000 en-
deavor. Finally, we would like to thank Drs. Brendan Allison, Febo Cincotti, and
Jeremy Hill for their tireless efforts to promote the dissemination of the software. In
addition, others have made important contributions to the project in different ways.
These are, in alphabetical order:

Erik Aarnoutse, Brendan Allison, Maria Laura Blefari, Sam Briskin, Simona Bufalari, Bob Cardillo, Nathaniel Elkins, Joshua Fialkoff, Emanuele Fiorilla, Dario
Gaetano, Christoph Guger of g.tec, Sebastian Halder, Jeremy Hill, Thilo Hinterberger, Jenny Hizver, Sam Inverso, Vaishali Kamat, Dean Krusienski, Marco Mat-
tiocco, Griffin "The Geek" Milsap, Yvan Pearson-Lecours, Robert Oostenveld,
Cristhian Potes, Christian Puzicha, Thomas Schreiner, Chintan Shah, Mark Span,
Chris Veigl, Janki Vora, Richard Wang, Shi Dong Zheng.

Sponsors

The BCI2000 is currently supported by a R01 grant from the NIH (NIBIB) to Gerwin Schalk. It was previously supported by a bioengineering research partnership
(BRP) grant from the NIH (NIBIB/NINDS) to Jonathan Wolpaw.

Albany, New York, USA   *Gerwin Schalk*
Tübingen, Germany   *Jürgen Mellinger*

References

1. Allison, B.Z., McFarland, D.J., Schalk, G., Zheng, S.D., Jackson, M.M., Wolpaw, J.R.: Towards an independent brain–computer interface using steady state visual evoked potentials.
   Clin. Neurophysiol. **119**(2), 399–408 (2008). doi:10.1016/j.clinph.2007.09.121
2. Bell, C.J., Shenoy, P., Chalodhorn, R., Rao, R.P.: Control of a humanoid robot by a noninvasive brain–computer interface in humans. J. Neural Eng. **5**(2), 214–220 (2008). doi:10.1088/
   1741-2560/5/2/012
3. Buch, E., Weber, C., Cohen, L.G., Braun, C., Dimyan, M.A., Ard, T., Mellinger, J., Caria, A.,
   Soekadar, S., Fourkas, A., Birbaumer, N.: Think to move: a neuromagnetic brain–computer
   interface (BCI) system for chronic stroke. Stroke **39**(3), 910–917 (2008). doi:10.1161/
   STROKEAHA.107.505313
4. Cabrera, A.F., Dremstrup, K.: Auditory and spatial navigation imagery in brain–computer interface using optimized wavelets. J. Neurosci. Methods **174**(1), 135–146 (2008). doi:10.1016/
   j.jneumeth.2008.06.026
5. Cincotti, F., Mattia, D., Aloise, F., Bufalari, S., Astolfi, L., De Vico Fallani, F., Tocci, A.,
   Bianchi, L., Marciani, M.G., Gao, S., Millan, J., Babiloni, F.: High-resolution EEG techniques for brain–computer interface applications. J. Neurosci. Methods **167**(1), 31–42 (2008).
   doi:10.1016/j.jneumeth.2007.06.031
6. Cincotti, F., Mattia, D., Aloise, F., Bufalari, S., Schalk, G., Oriolo, G., Cherubini, A., Marciani, M.G., Babiloni, F.: Non-invasive brain–computer interface system: towards its application as assistive technology. Brain Res. Bull. **75**(6), 796–803 (2008). doi:10.1016/
   j.brainresbull.2008.01.007
7. Farwell, L.A., Donchin, E.: Talking off the top of your head: toward a mental prosthesis utilizing event-related brain potentials. Electroencephalogr. Clin. Neurophysiol. **70**(6), 510–523
   (1988)
8. Felton, E.A., Wilson, J.A., Williams, J.C., Garell, P.C.: Electrocorticographically controlled
   brain–computer interfaces using motor and sensory imagery in patients with temporary subdural electrode implants. Report of four cases. J. Neurosurg. **106**(3), 495–500 (2007)

9. Kubánek, J., Miller, K.J., Ojemann, J.G., Wolpaw, J.R., Schalk, G.: Decoding flexion of individual fingers using electrocorticographic signals in humans. J. Neural Eng. **6**(6), 66,001–66,001 (2009). doi:10.1088/1741-2560/6/6/066001

10. Kübler, A., Nijboer, F., Mellinger, J., Vaughan, T.M., Pawelzik, H., Schalk, G., McFarland, D.J., Birbaumer, N., Wolpaw, J.R.: Patients with ALS can use sensorimotor rhythms to operate a brain–computer interface. Neurol. **64**(10), 1775–1777 (2005). doi:10.1212/01.WNL.0000158616.43002.6D

11. Leuthardt, E., Schalk, G., JR, J.W., Ojemann, J., Moran, D.: A brain–computer interface using electrocorticographic signals in humans. J. Neural Eng. **1**(2), 63–71 (2004)

12. Leuthardt, E., Miller, K., Schalk, G., Rao, R., Ojemann, J.: Electrocorticography-based brain computer interface – the Seattle experience. IEEE Trans. Neural Syst. Rehabil. Eng. **14**, 194–198 (2006)

13. Leuthardt, E., Miller, K., Anderson, N., Schalk, G., Dowling, J., Miller, J., Moran, D., Ojemann, J.: Electrocorticographic frequency alteration mapping: a clinical technique for mapping the motor cortex. Neurosurg. **60**, 260–270, discussion 270–271 (2007). doi:10.1227/01.NEU.0000255413.70807.6E

14. McFarland, D.J., Krusienski, D.J., Sarnacki, W.A., Wolpaw, J.R.: Emulation of computer mouse control with a noninvasive brain–computer interface. J. Neural Eng. **5**(2), 101–110 (2008). doi:10.1088/1741-2560/5/2/001. http://www.hubmed.org/display.cgi?uids=18367779

15. Mellinger, J., Schalk, G., Braun, C., Preissl, H., Rosenstiel, W., Birbaumer, N., Kübler, A.: An MEG-based brain–computer interface (BCI). NeuroImage **36**(3), 581–593 (2007). doi:10.1016/j.neuroimage.2007.03.019

16. Miller, K.J., Dennijs, M., Shenoy, P., Miller, J.W., Rao, R.P., Ojemann, J.G.: Real-time functional brain mapping using electrocorticography. NeuroImage **37**(2), 504–507 (2007). doi:10.1016/j.neuroimage.2007.05.029

17. Miller, K., Leuthardt, E., Schalk, G., Rao, R., Anderson, N., Moran, D., Miller, J., Ojemann, J.: Spectral changes in cortical surface potentials during motor movement. J. Neurosci. **27**, 2424–2432 (2007). doi:10.1523/JNEUROSCI.3886-06.2007. http://www.jneurosci.org/cgi/content/abstract/27/9/2424

18. Royer, A.S., He, B.: Goal selection versus process control in a brain–computer interface based on sensorimotor rhythms. J. Neural Eng. **6**(1), 16,005–16,005 (2009). doi:10.1088/1741-2560/6/1/016005

19. Schalk, G., McFarland, D., Hinterberger, T., Birbaumer, N., Wolpaw, J.: BCI2000: a general-purpose brain–computer interface (BCI) system. IEEE Trans. Biomed. Eng. **51**, 1034–1043 (2004)

20. Schalk, G., Kubánek, J., Miller, K.J., Anderson, N.R., Leuthardt, E.C., Ojemann, J.G., Limbrick, D., Moran, D., Gerhardt, L.A., Wolpaw, J.R.: Decoding two-dimensional movement trajectories using electrocorticographic signals in humans. J. Neural Eng. **4**(3), 264–275 (2007). doi:10.1088/1741-2560/4/3/012

21. Schalk, G., Leuthardt, E.C., Brunner, P., Ojemann, J.G., Gerhardt, L.A., Wolpaw, J.R.: Real-time detection of event-related brain activity. NeuroImage **43**(2), 245–249 (2008). doi:10.1016/j.neuroimage.2008.07.037

22. Schalk, G., Miller, K.J., Anderson, N.R., Wilson, J.A., Smyth, M.D., Ojemann, J.G., Moran, D.W., Wolpaw, J.R., Leuthardt, E.C.: Two-dimensional movement control using electrocorticographic signals in humans. J. Neural Eng. **5**(1), 75–84 (2008). doi:10.1088/1741-2560/5/1/008

23. Wilson, J., Felton, E., Garell, P., Schalk, G., Williams, J.: ECoG factors underlying multimodal control of a brain–computer interface. IEEE Trans. Neural Syst. Rehabil. Eng. **14**, 246–250 (2006)

24. Wisneski, K.J., Anderson, N., Schalk, G., Smyth, M., Moran, D., Leuthardt, E.C.: Unique cortical physiology associated with ipsilateral hand movements and neuroprosthetic implications. Stroke **39**(12), 3351–3359 (2008). doi:10.1161/STROKEAHA.108.518175

25. Wolpaw, J.R., McFarland, D.J.: Multichannel EEG-based brain–computer communication. Electroencephalogr. Clin. Neurophysiol. **90**(6), 444–449 (1994)

26. Wolpaw, J.R., McFarland, D.J.: Control of a two-dimensional movement signal by a noninvasive brain–computer interface in humans. Proc. Natl. Acad. Sci. USA **101**(51), 17,849–17,854 (2004). doi:10.1073/pnas.0403504101

27. Yamawaki, N., Wilke, C., Liu, Z., He, B.: An enhanced time–frequency-spatial approach for motor imagery classification. IEEE Trans. Neural Syst. Rehabil. Eng. **14**(2), 250–254 (2006)

Contents

Acronyms

ALS	Amyotrophic Lateral Sclerosis
AR	Autoregressive
BCI	Brain–Computer Interface
CAR	Common Average Reference
CSP	Common Spatial Patterns
CPU	Central Processing Unit
ECoG	Electrocorticogram
EEG	Electroencephalogram
EMG	Electromyogram
EOG	Electrooculogram
ERP	Evoked Response
FES	Functional electrical stimulation
FFT	Fast Fourier Transform
fMRI	Functional Magnetic Resonance Imaging
fNIR	Functional Near Infrared
ICA	Independent Component Analysis
IIR	Infinite Impulse Response
LDA	Linear Discriminant Analysis
MEG	Magnetoencephalography
MEM	Maximum Entropy Method
PET	Positron Emission Tomography
SCP	Slow Cortical Potentials
SSVEP	Steady-State Visual Evoked Potential
SWLDA	Stepwise Linear Discriminant Analysis
SVM	Support Vector Machine
TMS	Transcranial Magnetic Stimulation

Part I
User Guide

Chapter 1
Brain–Computer Interfaces

1.1 Introduction

Many patients are afflicted with neurological conditions or neurodegenerative diseases that disrupt the normal information flow from the brain to the spinal cord and eventually to the targets of that information, i.e., the muscles that effect the person's intent. Amyotrophic lateral sclerosis (ALS, or also called Lou Gehrig's disease), spinal cord injury, stroke, and many other conditions impair either the neural pathways controlling muscles, or impair the muscles themselves (Fig. 1.1-A). Those individuals that are most affected may lose all abilities to control muscles. Thus, they lose all options to communicate and become completely locked-in to their bodies. In absence of reversing the effects of the disorders, there are three principal options for restoring function.

The first option is to substitute the damaged neural pathways or muscles with pathways or muscles that are still functional (Fig. 1.1-B). While this substitution is often limited, it can still be useful. For example, patients can use eye movements to communicate [8, 9] or hand movements to produce synthetic speech (e.g., [2, 3, 16, 17]). The second option is to restore function by detecting nerve or muscle activity above the level of the injury (Fig. 1.1-C). For example, the Freehand prosthesis is restoring hand function to patients with spinal cord injuries [6, 11, 13]. The third option for restoring function is to provide the brain with a new and non-muscular output channel, a brain–computer interface (BCI), for conveying the user's intent to the external world (Fig. 1.1-D).

1.2 Brain–Computer Interfaces (BCIs)

A brain–computer interface (BCI) is a non-muscular communication system that a person can use to directly communicate his/her intent from the brain to the environment [31]. Thus, a BCI system attaches function to brain signals and thereby creates a new communication channel between the brain and a computer. The *language* of this communication is in part imposed on the brain (by the use of particular

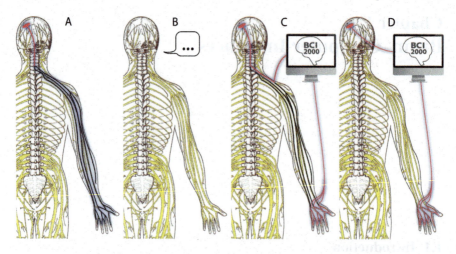

Fig. 1.1 Communication options for the paralyzed. **A**: Normal output communication channels from the brain to the periphery (e.g., the right hand) are disrupted. **B**: Option 1: Communication by substitution with other options (such as speech). **C**: Option 2: Communication by circumventing the impaired pathway. **D**: Option 3: Adding a new communication channel directly from the brain to an output device or an existing limb – a Brain–Computer Interface (BCI)

brain signal features that the BCI system extracts and uses for device control) and in part negotiated (by the continuous and mutual adaptations of both the user and the system).

Like any communication system, a BCI has an input (i.e., brain signals from the user), an output (i.e., device commands), components that translate the former into the latter, and an operating protocol that determines the onset, offset, and timing of operation. Thus, any BCI system can be described by four components: signal acquisition, which acquires signals from the brain; signal processing, which extracts signal features from brain signals and translates those into device commands; an output device, which acts upon these device commands and thereby effects the user's intent; and an operating protocol that guides operation (see Fig. 1.2). This implies that optimization of the BCI communication process is a complex and multidisciplinary problem that requires the appropriate integration of aspects of neuroscience, electrical engineering, computer science, and human factors.

Using different sensors and brain signals, many studies over the past two decades have evaluated the possibility that BCI systems could provide new augmentative technology that does not require muscle control (e.g., [1, 4, 5, 10, 12, 14, 15, 18, 20, 21, 23–31]). These brain–computer interface (BCI) systems measure specific features of brain activity and translate them into device control signals. Thus, a BCI system derives and utilizes control signals to effect the user's intent, and it usually does so by allowing the user to make a selection. This selection capacity is often realized using a computer cursor (e.g., [10, 29]), but also in other ways such as controlling an arrow on a dial [20], a moving robot [19], or by controlling other external devices [4, 22]. The key performance characteristics of BCI systems are

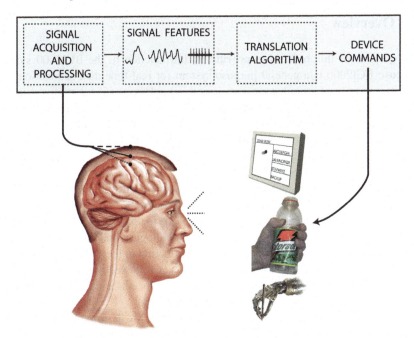

Fig. 1.2 Basic design and operation of any BCI system. Signals from the brain can be acquired by electrodes on the scalp, the cortical surface, or from within the cortex, and processed to extract specific signal features (e.g., time- or frequency-domain measurements) that reflect the user's intent. These features are translated into commands that operate a device (e.g., a simple word processing program, a wheelchair, or a neuroprosthesis)

speed (i.e., how long it takes to make a selection) and precision (i.e., the fraction of executed selections that are consistent with the user's intent). Current systems allow for one selection within several seconds at a relatively high accuracy (e.g., 90% accuracy in a binary task). Expressed in bit rate, which combines both speed and accuracy, the sustained performance of typical BCI systems is still modest. This bit rate typically ranges between 5–25 bits/min [31], although some studies have reported rates that were higher than that [7].

In consequence, a key problem in the translation of the BCI demonstrations mentioned above into clinically and commercially successful communication systems for the disabled is to increase the modest performance of current BCI systems. Addressing this problem requires comprehensive evaluations of different BCI methodologies. This process is greatly facilitated by a general-purpose BCI platform that can readily support any BCI methodology and that facilitates interchange of data and experimental protocols. This need for a general-purpose BCI platform provided the impetus for the development of the BCI2000 system that is described in this book.

1.3 Overview

The purpose of this book is to provide an introduction to the BCI2000 system. Because BCI2000 is a general-purpose system for real-time data acquisition, signal processing, and stimulus presentation, this book is meant for students and researchers who would like to implement real-time systems. Because BCI2000 is *not* a signal processing toolbox, and *not* simply a collection of BCI demos, but rather a robust and capable framework for a wide range of experiments, the BCI2000 system generally and this book specifically will be most useful to people who have a serious interest in research in this area.

Because as of yet there are very few comprehensive educational curricula or other general resources on BCI research, we felt that it would be important to include several chapters on basic aspects of BCI systems. This includes a discussion of relevant sensors, relevant brain signals, recording methodologies, and BCI signal processing. This background sets the stage for an introduction to the BCI2000 system, and a tour through its most important features. This is followed by user and programming tutorials, as well as a discussion of advanced concepts and exercises. The book is completed with a comprehensive technical reference.

We use several text conventions in this book when we describe BCI2000 or give step-by-step instructions. These conventions are described briefly here.

- Menu items in BCI2000 or other programs are shown as *File*.
- Button text in BCI2000 or other programs is also shown as *Start*.
- Parameter names (e.g., within the BCI2000 configuration window) appear as **SubjectName**.
- *Displayed* messages appear on the screen during testing.
- Website URLs, such as www.bci2000.org, are shown underlined.
- File locations appear as `c:/BCI2000/prog/operat.exe`.
- C++ code segments appear as `cout << "Hello World." << endl;`

In several places in the book, the reader is provided with step-by-step instructions to complete a tutorial. These instructions appear in a text box with numbered instructions, such as

---

Example Instructions

1. Start BCI2000.
2. Press *Config*.
3. ...

References

1. Birbaumer, N., Ghanayim, N., Hinterberger, T., Iversen, I., Kotchoubey, B., Kübler, A., Perelmouter, J., Taub, E., Flor, H.: A spelling device for the paralysed. Nature **398**(6725), 297–298 (1999)

2. Chen, Y.L., Tang, F.T., Chang, W.H., Wong, M.K., Shih, Y.Y., Kuo, T.S.: The new design of an infrared-controlled human–computer interface for the disabled. IEEE Trans. Rehabil. Eng. **7**, 474–481 (1999)
3. Damper, R.I., Burnett, J.W., Gray, P.W., Straus, L.P., Symes, R.A.: Hand-held text-to-speech device for the non-vocal disabled. J. Biomed. Eng. **9**, 332–340 (1987)
4. Donoghue, J.P., Nurmikko, A., Black, M., Hochberg, L.R.: Assistive technology and robotic control using motor cortex ensemble-based neural interface systems in humans with tetraplegia. J. Physiol. **579**(3), 603–611 (2007). doi:10.1113/jphysiol.2006.127209
5. Farwell, L.A., Donchin, E.: Talking off the top of your head: toward a mental prosthesis utilizing event-related brain potentials. Electroencephalogr. Clin. Neurophysiol. **70**(6), 510–523 (1988)
6. Ferguson, K.A., Polando, G., Kobetic, R., Triolo, R.J., Marsolais, E.B.: Walking with a hybrid orthosis system. Spinal Cord **37**, 800–804 (1999)
7. Gao, X., Xu, D., Cheng, M., Gao, S.: A BCI-based environmental controller for the motion-disabled. IEEE Trans. Neural Syst. Rehabil. Eng. **11**(2), 137–140 (2003). doi:10.1109/TNSRE.2003.814449
8. Gerhardt, L., Sabolcik, R.: Eye tracking apparatus and method employing grayscale threshold values. US Patent 5,481,622, 1996
9. Grauman, K., Betke, M., Gips, J., Bradski, G.: Communication via eye blinks – detection and duration analysis in real time. In: 2001 IEEE Computer Society Conference on Computer Vision and Pattern Recognition, pp. 1010–1017. IEEE Comput. Soc., Los Alamitos (2001)
10. Hochberg, L.R., Serruya, M.D., Friehs, G.M., Mukand, J.A., Saleh, M., Caplan, A.H., Branner, A., Chen, D., Penn, R.D., Donoghue, J.P.: Neuronal ensemble control of prosthetic devices by a human with tetraplegia. Nature **442**(7099), 164–171 (2006). doi:10.1038/nature04970
11. Hoffer, J.A., Stein, R.B., Haugland, M.K., Sinkjaer, T., Durfee, W.K., Schwartz, A.B., Loeb, G.E., Kantor, C.: Neural signals for command control and feedback in functional neuromuscular stimulation: a review. J. Rehabil. Res. Dev. **33**, 145–157 (1996)
12. Kennedy, P.R., Bakay, R.A., Moore, M.M., Goldwaithe, J.: Direct control of a computer from the human central nervous system. IEEE Trans. Rehabil. Eng. **8**(2), 198–202 (2000)
13. Kilgore, K.L., Peckham, P.H., Keith, M.W., Thrope, G.B., Wuolle, K.S., Bryden, A.M., Hart, R.L.: An implanted upper-extremity neuroprothesis: follow-up of five patients. J. Bone Jt. Surg. **79-A**, 533–541 (1997)
14. Kübler, A., Kotchoubey, B., Hinterberger, T., Ghanayim, N., Perelmouter, J., Schauer, M., Fritsch, C., Taub, E., Birbaumer, N.: The Thought Translation Device: a neurophysiological approach to communication in total motor paralysis. Exp. Brain Res. **124**(2), 223–232 (1999)
15. Kübler, A., Nijboer, F., Mellinger, J., Vaughan, T.M., Pawelzik, H., Schalk, G., McFarland, D.J., Birbaumer, N., Wolpaw, J.R.: Patients with ALS can use sensorimotor rhythms to operate a brain–computer interface. Neurol. **64**(10), 1775–1777 (2005). doi:10.1212/01.WNL.0000158616.43002.6D
16. Kubota, M., Sakakihara, Y., Uchiyama, Y., Nara, A., Nagata, T., Nitta, H., Ishimoto, K., Oka, A., Horio, K., Yanagisawa, M.: New ocular movement detector system as a communication tool in ventilator-assisted Werdnig–Hoffmann disease. Dev. Med. Child Neurol. **42**, 61–64 (2000)
17. LaCourse, J.R., Hludik, F.C. Jr.: An eye movement communication–control system for the disabled. IEEE Trans. Biomed. Eng. **37**, 1215–1220 (1990)
18. McFarland, D.J., Neat, G.W., Wolpaw, J.R.: An EEG-based method for graded cursor control. Psychobiol. **21**, 77–81 (1993)
19. Millán, J. del R., Renkens, F., Mouriño, J., Gerstner, W.: Noninvasive brain-actuated control of a mobile robot by human EEG. IEEE Trans. Biomed. Eng. **51**(6), 1026–1033 (2004)
20. Müller, K., Blankertz, B.: Toward noninvasive brain–computer interfaces. IEEE Signal Process. Mag. **23**(5), 126–128 (2006)
21. Pfurtscheller, G., Flotzinger, D., Kalcher, J.: Brain–computer interface – a new communication device for handicapped persons. J. Microcomput. Appl. **16**, 293–299 (1993)
22. Pfurtscheller, G., Guger, C., Müller, G., Krausz, G., Neuper, C.: Brain oscillations control hand orthosis in a tetraplegic. Neurosci. Lett. **292**(3), 211–214 (2000)

23. Santhanam, G., Ryu, S.I., Yu, B.M., Afshar, A., Shenoy, K.V.: A high-performance brain–computer interface. Nature **442**(7099), 195–198 (2006). doi:10.1038/nature04968
24. Serruya, M., Hatsopoulos, N., Paninski, L., Fellows, M., Donoghue, J.: Instant neural control of a movement signal. Nature **416**(6877), 141–142 (2002)
25. Sutter, E.E.: The brain response interface: communication through visually guided electrical brain responses. J. Microcomput. Appl. **15**, 31–45 (1992)
26. Taylor, D.M., Tillery, S.I., Schwartz, A.B.: Direct cortical control of 3D neuroprosthetic devices. Science **296**, 1829–1832 (2002)
27. Vaughan, T.M., McFarland, D.J., Schalk, G., Sarnacki, W.A., Krusienski, D.J., Sellers, E.W., Wolpaw, J.R.: The Wadsworth BCI research and development program: at home with BCI. IEEE Trans. Neural Syst. Rehabil. Eng. **14**(2), 229–233 (2006)
28. Wessberg, J., Stambaugh, C.R., Kralik, J.D., Beck, P.D., Laubach, M., Chapin, J.K., Kim, J., Biggs, S.J., Srinivasan, M.A., Nicolelis, M.A.: Real-time prediction of hand trajectory by ensembles of cortical neurons in primates. Nature **408**, 361–365 (2000)
29. Wolpaw, J., McFarland, D.: Control of a two-dimensional movement signal by a non-invasive brain–computer interface in humans. Proc. Natl. Acad. Sci. USA **101**, 17849–17854 (2004)
30. Wolpaw, J.R., McFarland, D.J., Neat, G.W., Forneris, C.A.: An EEG-based brain–computer interface for cursor control. Electroencephalogr. Clin. Neurophysiol. **78**(3), 252–259 (1991)
31. Wolpaw, J.R., Birbaumer, N., McFarland, D.J., Pfurtscheller, G., Vaughan, T.M.: Brain–computer interfaces for communication and control. Electroencephalogr. Clin. Neurophysiol. **113**(6), 767–791 (2002)

Chapter 2
Brain Sensors and Signals

2.1 Relevant Sensors

A variety of sensors for monitoring brain activity exist, and could in principle provide the basis for a BCI. These include, besides electroencephalography (EEG) and more invasive electrophysiological methods such as electrocorticography (ECoG) and recordings from individual neurons within the brain, magnetoencephalography (MEG), positron emission tomography (PET), functional magnetic resonance imaging (fMRI), and optical imaging (i.e., functional Near InfraRed (fNIR)). However, MEG, PET, fMRI, and fNIR are still technically demanding and expensive, which impedes widespread use. Despite these impediments, several studies have recently explored the value of these modalities for BCI research [10, 11, 42, 60, 82, 97, 98, 108–110, 118]. Furthermore, PET, fMRI, and fNIR, which depend on metabolic processes, have long time constants and thus seem to be less amenable to rapid communication. At present, non-invasive and invasive electrophysiological methods (i.e., EEG, ECoG, and single-neuron recordings, see illustration in Fig. 2.1) are the only methods that use relatively simple and inexpensive equipment and have high temporal resolution. Thus, these three alternatives are at present the only methods that can offer the possibility of a new non-muscular communication and control channel – a practical brain–computer interface.

The first and least invasive alternative uses EEG, which is recorded from the scalp [6, 22, 39, 40, 54, 59, 61, 64, 73, 76, 101, 106, 114, 116, 117]. These BCIs support much higher performance than previously assumed, including two- and three-dimensional cursor movement [59, 66, 114]. However, the acquisition of such high levels of control typically requires extensive user training. Furthermore, EEG has low spatial resolution, which will eventually limit the amount of information that can be extracted, and it is also susceptible to artifacts from other sources.

The second alternative uses ECoG, which is recorded from the cortical surface [23, 46, 47, 111]. It has higher spatial resolution than EEG (i.e., tenths of millimeters vs. centimeters [25]), broader bandwidth (i.e., 0–500 Hz [99] vs. 0–50 Hz), higher characteristic amplitude (i.e., 50–100 μV vs. 10–20 μV), and far less vulnerability to artifacts such as EMG [3, 25] or ambient noise. While this method is invasive,

Fig. 2.1 Different types of sensors most commonly used in BCI research. *A*: Electrodes are placed non-invasively on the scalp (electroencephalography (EEG)). *B*: Electrodes are placed on the surface of the brain (electrocorticography (ECoG)). *C*: Electrodes are placed invasively within the brain (single-neuron recordings). (From [112])

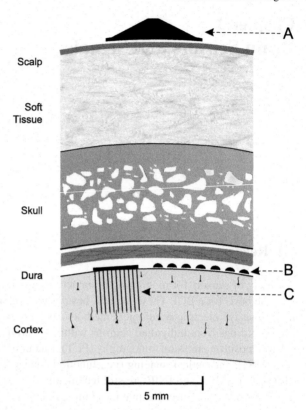

the use of these electrodes that do not penetrate the cortex may combine excellent signal fidelity with good long-term stability [7, 49, 52, 119].

The third and most invasive alternative uses microelectrodes to measure local activity (i.e., action or field potentials) from multiple neurons within the brain [21, 32, 45, 67, 84, 92, 95, 103]. Signals recorded within cortex have higher fidelity and might support BCI systems that require less training than EEG-based systems. However, clinical implementations of intracortical BCIs are currently impeded mainly by the difficulties in maintaining stable long-term recordings [20, 93, 100], by the substantial technical requirements of single-neuron recordings, and by the need for continued intensive expert oversight. For these reasons, practically all BCI demonstrations in humans to date have been achieved with, and the examples in this book are using or meant for, EEG or ECoG recordings.

2.2 Brain Signals and Features

2.2.1 Using Brain Signals for Communication

Successful creation of a new communication channel – directly from the brain to an output device – depends on two requirements. The first requirement is the use of an adequate sensor that can effectively measure the brain signal features that can communicate a user's intent. As described in the previous section, multiple sensors exist that can in principle detect relevant signals. Practicality and speed considerations exclude most of these options, so that almost all BCI systems to date depend on detection of electrophysiological signals using sensors on the scalp, on the surface of the brain, or within the brain. In humans, safety and/or stability issues have confined most studies to electroencephalographic (EEG) recordings from the scalp. The second requirement is the definition and negotiation of a mutual language (i.e., brain signal features such as time-domain or frequency-domain measurements at particular locations), so that, as in any other communication system, the user may use the symbols of this language to communicate intent, and the computer can detect these symbols and effect this intent.

For two reasons, the language of BCI communication cannot be completely arbitrary. First, the brain might simply not be physically able to produce the symbols of this language. For example, one might define the arbitrary language as the amplitude coherence between two different frequency bands at one particular location, and its symbols could be discrete coherence amplitudes, but the brain might simply not be physically able to produce changes in coherence amplitude at the selected frequencies and locations. Second, the brain might be able to produce the symbols of this language, but might not be able to use them to convey intent. For example, one might define the arbitrary language as amplitude modulations at 10 Hz over visual areas of the brain. Many studies have shown that repetitive visual stimuli at particular frequencies (such as 10 Hz) can evoke oscillatory responses in the brain [63], so clearly the brain is physically able to modulate activity at 10 Hz and can thus produce different symbols of that arbitrary language. However, it might not be able to produce these symbols without the visual stimuli, or might not be able to use these symbols to convey intent.

In summary, there is no theoretical basis for selecting the language (i.e., brain signal) that is most useful for BCI communication. Furthermore, any clinically successful BCI will necessarily be under the influence of practical considerations such as risk, benefit, and price. Thus, it is currently unclear which brain signal and which sensor modality (EEG, ECoG, or single neuron recordings) will ultimately be most beneficial given these constraints. At the same time, experimental evidence is able to provide some guidance on which brain signals to utilize for BCI communication. For example, many studies have shown that particular imagined tasks (e.g., hand movements) have detectable effects on particular brain signals. Taking advantage of this phenomenon, people can communicate simple messages using imagery of hand movements. Other studies have shown that the presentation of desired stimuli produces detectable brain signal responses. By presenting multiple stimuli and by detecting the response to the desired stimulus, people can communicate which item they desire. These two possibilities are representative for the phenomena most relevant for BCI communication in humans, and are described in more detail in the following two sections.

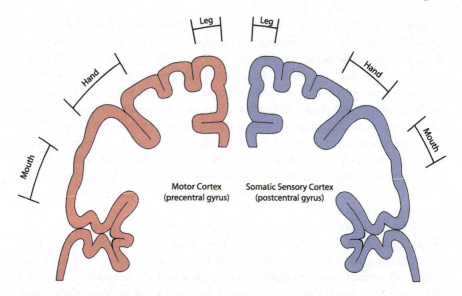

Fig. 2.2 The brain figure on the top shows a vertical cross-section along motor (*left*) and sensory (*right*) cortical areas. The motor cortex is displayed in *red*. Particular areas in the motor cortex are associated with function of particular limbs (i.e., "motor homunculus"). Similarly, sensory cortex is shown in *blue*. Particular areas in sensory cortex are also associated with sensory function of different limbs (i.e., "sensory homunculus")

2.2.2 Mu/Beta Oscillations and Gamma Activity

Most people exhibit prominent oscillations in the 8–12 Hz band of the EEG recorded over sensorimotor areas (see Fig. 2.2) when they are not actively engaged in motor action, sensory processing, or in imaginations of such actions or processing [24, 27, 37] (reviewed in [69]). This oscillation is usually called mu rhythm and is thought to be produced by thalamocortical circuits [69]. The lack of modern acquisition and processing methodologies have initially not made it possible to detect the mu rhythm in many people [8], but computer-based analyses have since revealed that the mu rhythm is in fact present in a large majority of people [70, 71]. Such analyses have also demonstrated that the mu rhythm is usually associated with 18–25 Hz beta rhythms. While some of these beta rhythms are harmonics of mu rhythms, some are separable by topography and/or timing from mu rhythms, and thus at least appear to be independent EEG features [57, 70, 71].

Because mu/beta rhythm changes are associated with normal motor/sensory function, they could be good signal features for BCI-based communication. Movement or preparation for movement, but typically not specific aspects of movements such as its direction [104], are typically accompanied by a decrease in mu and beta activity over sensorimotor cortex, particularly contralateral to the movement. Furthermore, mu/beta rhythm changes also occur with motor imagery (i.e., imagined movement) [57, 72]. Because people can change these rhythms without engaging

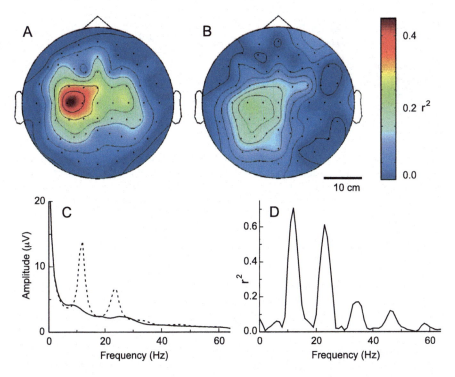

Fig. 2.3 Examples of mu/beta rhythm signals (modified from [85]). **A, B**: Topographical distribution on the scalp of the difference (measured as r^2 (the proportion of the single-trial variance that is due to the task)), calculated for actual (**A**) and imagined (**B**) right-hand movements and rest for a 3-Hz band centered at 12 Hz. **C**: Example voltage spectra for a different subject and a location over left sensorimotor cortex (i.e., C3 (see [94])) for comparing rest (*dashed line*) and imagery (*solid line*). **D**: Corresponding r^2 spectrum for rest vs. imagery. Signal modulation is focused over sensorimotor cortex and in the mu- and beta-rhythm frequency bands

in actual movements, these rhythms could serve as the basis for a BCI. Figure 2.3 shows the basic phenomenon of mu/beta-rhythm modulations in the EEG.

Similar to EEG, activity in the mu/beta bands recorded using ECoG also decreases with motor tasks [12, 14, 28, 48, 62, 77, 96]. In addition, activity in the gamma range (i.e., >40 Hz) has been found to increase with these tasks [13, 47, 48, 62]. With isolated exceptions (e.g., [74]), task-related changes in these higher frequencies have not been reported in the EEG. There are indications that gamma activity reflects activity of local neuronal groups [41, 62], and thus could be most directly reflective of specific details of movement. Indeed, recent studies have shown relationships of gamma activity with specific kinematic parameters of hand movements [9, 79, 83, 86].

In summary, many studies have shown using EEG [36, 54, 61, 64, 73, 113–116] or ECoG [23, 46, 47, 87, 111] that humans can use motor imagery to modulate activity in the mu, beta, or gamma bands, and to thereby control a BCI system.

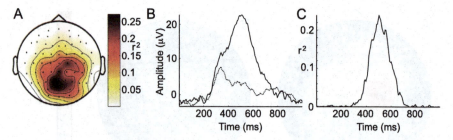

Fig. 2.4 Example characteristics of the P300 response (data courtesy of Dr. Eric Sellers, Wadsworth Center/East Tennessee State University). *Left*: Topographical distribution of the P300 potential at 500 ms after stimuli, measured as r^2 and calculated between desired and not desired stimuli. *Center*: The time courses at electrode location Pz of the voltages for desired (*solid line*) or not desired (*dashed line*) stimuli. *Right*: Corresponding r^2 time course

2.2.3 The P300 Evoked Potential

In addition to brain responses modulated by motor action or motor imagery, evoked potentials may also be useful for BCI operation. For example, numerous studies over the past four decades have shown that presentation of infrequent stimuli typically evokes a positive response (called the "P300" or "oddball" potential) in the EEG over parietal cortex about 300 ms after stimulus presentation (see [17, 102, 107]; [15, 18, 80] for review; Fig. 2.4). The amplitude of the P300 potential is largest at the parietal electrode sites and is attenuated as the recording sites move to central and frontal locations [15]. A P300 is usually elicited if four conditions are met. First, a random sequence of stimulus events must be presented. Second, a classification rule that separates the series of events into two categories must be applied. Third, the user's task must require using the rule. Fourth, one category of events must be presented infrequently [16].

Using experimental paradigms that implement these four conditions, the P300 potential has been used as the basis for a BCI system in many studies [1, 5, 19, 22, 33, 68, 78, 88–91, 106]. The classical format developed by Donchin and colleagues [22] presents the user with a matrix of characters (Fig. 2.5). The rows and columns in this matrix flash successively and randomly at a rapid rate (e.g., eight flashes per second). The user selects a character by focusing attention on it and counting

Fig. 2.5 The classical P300-based spelling paradigm developed by Donchin [19, 22]. Rows and columns of the matrix flash in a block-randomized fashion. The row or column that contains the desired character evokes a P300 potential

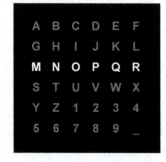

how many times it flashes. The row or column that contains this character evokes a P300 response, whereas all others do not. After averaging several responses, the computer can determine the desired row and column (i.e., the row/column with the highest P300 amplitude), and thus the desired character.

2.3 Recording EEG

2.3.1 Introduction

After discussing the brain signals most commonly used for BCI operation, this section describes the relevant principles of brain signal recordings and the types of signal artifacts that are typically encountered. These descriptions are mostly focused on EEG. The same general recording principles also apply to ECoG recordings. While most types of artifacts are typically encountered only with EEG, artifacts can also be detected in ECoG [3]. Brain signals are detected using different types of metal electrodes that are placed on the scalp (EEG) or on the surface of the brain (ECoG). These electrodes measure small electrical potentials that reflect the activity of neurons within the brain. To detect the tiny amplitude of these signals, they first have to be amplified. Any biosignal amplifier measures the potential difference (i.e., the voltage) between two electrodes. In most BCI systems, the second of these two electrodes is always the same, i.e., measurements are "unipolar" rather than "bipolar." In other words, measurements from all electrodes are referred to one common electrode, which consequently is called "reference," and typically labeled *Ref*. To improve signal quality, amplifiers require the connection of a "ground" electrode, which is typically labeled *Gnd*.

EEG electrodes are small metal plates that are attached to the scalp using a conducting electrode gel. They can be made from various materials. Most frequently, tin electrodes are used, but there are gold, platinum, and silver/silver-chloride (Ag/AgCl) electrodes as well. Tin electrodes are relatively inexpensive and work well for the typical BCI-related application. At the same time, tin electrodes introduce low-frequency drifting noise below 1 Hz, which makes them unsuitable for some applications (e.g., Slow Cortical Potential measurements or low-noise evoked potential recordings).

An important but often neglected detail: mixing electrodes made from different materials in the same recording will result in DC voltage offsets between electrodes. These offsets are due to electrochemical contact potentials and can often be larger in amplitude than what a typical amplifier can tolerate. This will result in a greatly diminished signal-to-noise ratio. Thus, you should always use electrodes made from the same material in a particular recording.

2.3.2 Electrode Naming and Positioning

The standard naming and positioning scheme for EEG applications is called the 10–20 international system [35]. It is based on an iterative subdivision of arcs on the scalp starting from particular reference points on the skull: Nasion (Ns), Inion (In), and Left and Right Pre-Auricular points (PAL and PAR, respectively). The intersection of the longitudinal (Ns–In) and lateral (PAL–PAR) diagonals is named the Vertex, see Fig. 2.6-A. The 10–20 system originally included only 19 electrodes (Fig. 2.6-B, [35]). This standard was subsequently extended to more than 70 electrodes (Fig. 2.6-C, [94]). This extension also renamed electrodes T3, T5, T4, and T6, into T7, P7, T8, and P8, respectively. Sometimes, one of the electrodes mounted in these positions is used as reference electrode. More often, the ear lobe or mastoid (i.e., bony outgrowth behind the ear) are used. For example, a typical recording may have the ground electrode placed on the mastoid and the reference on the ear lobe on the opposite side.

Acquiring EEG from more than a single location is necessary to be able to determine the optimum location for the BCI purpose. It also greatly facilitates the identification of signal artifacts. Thus, for research purposes we strongly recommend to record from as many locations as possible – at least from 16. For clinical applications, we suggest to initially record from at least 16 locations. Once effective BCI operation has been established and the optimal locations have been determined, the electrode montage can be optimized to record from fewer locations.

2.3.3 Important Brain Areas and Landmarks for BCIs

The brain consists of several distinct areas and landmarks. The approximate location of these areas and landmarks can be determined from the extended 10–20 system shown in Fig. 2.6-C. One of these landmarks is the central sulcus, which is also called Rolandic fissure. The central sulcus runs approximately along the lines between electrodes CPz–C2–C4 and CPz–C1–C3, respectively. On each hemisphere, the central sulcus divides the brain into the frontal lobe (in frontal direction, i.e., toward the nose) and parietal lobe (in posterior direction, i.e., toward the back of the head). The frontal lobe contains, among other areas, the primary motor cortex, i.e., the area of the brain that is most immediately involved in the execution of movements. The parietal lobe contains, among other areas, primary sensory cortex, i.e., the area of the brain that is a direct neighbor of the primary motor cortex and that is most directly involved in processing sensory information from different body parts. Another important landmark is the Sylvian fissure, which is also called lateral sulcus. It runs along the lines connecting CP6–C6–FT8–FT10 and CP5–C5–FT7–FT9, respectively. It separates the temporal lobe, an area in the brain responsible for auditory processing and memory, from the frontal and parietal lobes.

2.3.4 Placing Electrodes with a Cap

Accurate placement of many electrodes on the scalp is time consuming and requires practice. EEG caps greatly facilitate this process. These caps are made of elastic

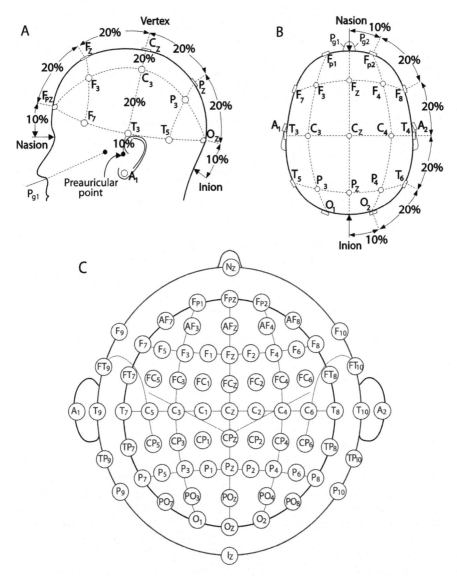

Fig. 2.6 Electrodes on the original 10–20 scheme (figures **A** and **B**, re-drawn from [35]). Current extensions to this scheme define more than 70 locations (figure **C**, re-drawn from [94])

fabric (often available in different sizes), and electrodes are already fixed in the proper configuration. One proven technique to place electrodes using such caps is the following:

- Mark the vertex on the subject's scalp using a felt-tip pen or some other similar method. Begin by locating the nasion and inion on the subject as indicated in panel A of Fig. 2.6. Using a tape measure, find the distance between these two locations. The point half-way between the two points is the vertex. Make a mark at that point for later reference. (Other 10–20 points could be located in a similar manner.)

- Mark scalp positions for Fpz and Oz. The Fpz position is above the nasion 10% of the distance from the nasion to the inion. The Oz position is above the inion the same distance.
- Identify the Cz electrode on the EEG cap and place the cap to position the Cz electrode on the vertex.
- Keeping Cz fixed, slide the cap onto the head.
- While ensuring that Cz does not shift, adjust the cap such that the Fz–Cz–Pz line is on the midline; Fp1–Fp2 line is horizontal, and at the level of the Fpz mark; the O1–O2 line is horizontal, and at the level of the Oz mark.
- You can now fix *Ref* and *Gnd* electrodes. These electrodes are attached in one of a few typical configurations. One common configuration is to attach the Ref electrode to one earlobe, and the Gnd electrode to the mastoid on the same side of the head. Another possible configuration is to attach Ref to one mastoid and Gnd to the other mastoid. This choice is influenced by the used cap technology, which may have separate electrodes outside the cap for reference and ground, or may have these electrodes embedded in the cap directly.

2.3.5 Removing Artifacts and Noise Sources

2.3.5.1 Introduction

Electrical power lines use sinusoidal voltages with a frequency of 50 or 60 Hz, depending on your country. Generally, 50 Hz are used in Europe, Asia, Africa, and parts of South America; 60 Hz are used in North America, and parts of South America. Voltages are typically 110 or 230 volts, and thus exceed the EEG's 50 to 100 microvolts by a factor of 2×10^6, or 126 dB. Therefore, mains interference is ubiquitous in EEG recordings, especially if taken outside specially equipped, shielded rooms. Most EEG amplifiers provide a so-called notch filter that suppresses signals in a narrow band around the power line frequency.

Amplifier notch filters are designed to suppress a certain amount of mains interference. When there is mains interference still visible in the signal after activating the amplifier's notch filter, this is often due to high electrode impedance (Fig. 2.7).

2.3.5.2 Artifacts due to Eye Blinks

Eye blink artifacts are generated by fast movements of the eyelid along the cornea, such as during an eye blink. By friction between lid and cornea, this movement results in charge separation, with a dominantly dipolar charge distribution, and the

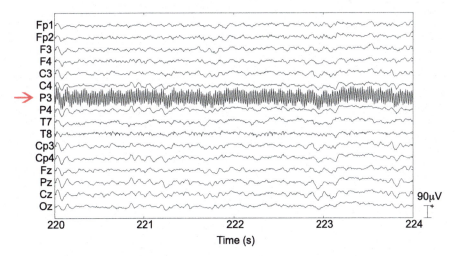

Fig. 2.7 Artifacts due to power line interference. This figure shows an example for one signal channel (marked by the *arrow*) that is contaminated by regular high frequency (i.e., 60 Hz) noise

dipole moment pointing in up-down-direction. In the EEG, this effect is recorded as a positive peak that lasts a few tenths of a second, is most prominent in the frontopolar region, but propagates to all the electrodes of the montage, attenuating with distance from the front (Fig. 2.8).

The frequency content of eye blink artifacts is negligible in the alpha band (around 10 Hz), so they have no strong effect on the use of sensorimotor rhythms. At the same time, their time-domain amplitude is quite large so that analyses in the

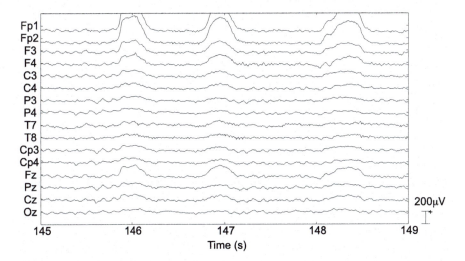

Fig. 2.8 Eye blink artifacts. This figure shows examples for signal contamination by eye blink artifacts. These artifacts are most prominent in frontal channels (channels *Fp1* and *Fp2*), but have effects on all channels

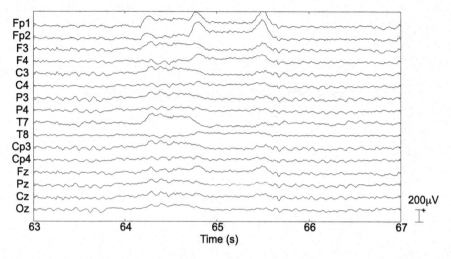

Fig. 2.9 Eye movement artifacts. Eye movements have a prominent effect on many channels. See the positive or negative deflections during early, middle and late parts of this recording

time domain (such as averaged P300 wave forms) can be strongly influenced by their presence.

2.3.5.3 Artifacts due to Eye Movements

Ocular artifacts (electrooculographic (EOG) signals) are produced by eye movements, and generated by a frictive mechanism that is similar to the one described above for eye blinks, except that it involves the retina and cornea rather than cornea alone. The effect on frontopolar and frontotemporal electrodes can be symmetric or asymmetric, depending whether the movement is vertical or horizontal, respectively. The effect of eye movement artifacts on frequency- or time-domain analysis is quite similar to that of blink artifacts, except that their frequency content is even lower, and amplitudes tend to be larger (Fig. 2.9).

2.3.5.4 Artifacts due to Muscle Movements

Muscular artifacts (electromyographic (EMG) signals, Fig. 2.10) must be carefully checked at the beginning of each recording, and verified throughout the recording. This is because the frequency distribution of EMG signals is very broad, so that they have a profound effect on amplitudes in typical mu/beta frequency ranges. The most common sources of EMG are the muscles that lift the eye brows, and those that close the jaw. Both groups of muscles can be inadvertently contracted during an experimental session. Keeping the mouth slightly open (or the tip of the tongue between the foreteeth) is a good strategy to avoid jaw-generated EMG.

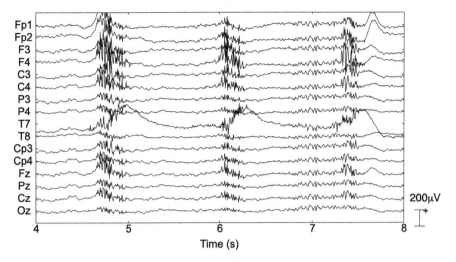

Fig. 2.10 Muscle movement artifacts. Many channels exhibit high frequency noise; in contrast to the power line interference, this noise is highly variable over time

2.4 BCI Signal Processing

The previous section described the two frequency- and time-domain phenomena most relevant to human BCI research, i.e., mu/beta rhythms and gamma activity, and the P300 evoked potential, respectively. Many studies have shown how these phenomena can be extracted and translated into device commands using different methods. All currently used procedures are listed in recent review articles on BCI feature extraction and translation methods [4, 50, 58]. The following sections describe an analysis approach that can realize many of these techniques. This approach is implemented in the example configurations of the BCI2000 software.

2.4.1 Introduction

BCI signal processing is a difficult problem. In addition to typical problems faced by any communication system (e.g., signals are contaminated with noise during transmission), it is initially and even during subsequent operation not clear which brain signals actually carry the message the user wants to communicate. In other words, it is the task of BCI signal processing to decode a message in a language we do not know much about. Fortunately, experimental evidence can provide some basic guidance. This guidance comes from observations that particular tasks (such as imagined hand movements) have particular effects on specific brain signals (such as the mu rhythm measured at a particular location). Even with this information, the choice of signals and tasks is still difficult, because it is likely that it is suboptimal (so that a completely different signal and task might provide improved performance) and

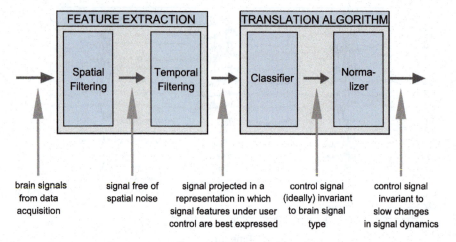

Fig. 2.11 Signal processing model utilized in BCI2000. This model consists of feature extraction and translation and can describe all common BCI methods

because it has to be optimized for each individual. In other words, even when only considering one possible physiological signal (such as the mu rhythm), the imagery task, best frequencies, and best locations have to be selected for each individual. The difficulty of choosing signals and tasks could be regarded as the *signal identification problem* of BCI communication. Assuming that a good candidate signal (e.g., the signal amplitude at a particular frequency or location) has been identified, traditional BCI signal processing usually employs two components to translate these signals into device commands: feature extraction and the translation algorithm.[1]

Feature extraction consists of two procedures: spatial filtering and temporal filtering (Fig. 2.11). Each of these procedures may have different realizations. The following paragraphs describe those realizations that are relevant to sensorimotor rhythm and P300 processing.

2.4.2 Spatial Filtering

The initial step in feature extraction is the application of a spatial filter that may have many possible realizations. The purpose of the spatial filter is to reduce the effect of spatial blurring. Spatial blurring occurs as an effect of the distance between the sensor and the signal sources in the brain, and because of the inhomogeneities of the tissues in between. Different approaches to spatial filtering have attempted

---

[1] We use the term *translation algorithm* instead of *classification* throughout this book, because the typically continuous nature of the device control signals produced by BCI signal processing is better expressed by the term translation algorithm rather than the discrete output that is typically implied by the term classification.

to reduce this blurring, and thus to increase signal fidelity. The most sophisticated approaches attempt to deblur signals using a realistic biophysical head model that is optimized for each user and whose parameters are derived from various sources such as magnetic resonance imaging (MRI) (e.g., [44]). While these approaches do increase signal quality in carefully controlled experiments, they are currently impractical for most experiments and for clinical applications. Other approaches do not require external parameterization of a complex model, but rather are simply driven by the signals that are submitted to it. For example, Independent Component Analysis (ICA) has been used to decompose brain signals into statistically independent components [51], which can be used to get a more effective signal representation. (This approach is called blind deconvolution in microscopy applications (e.g., [105]).) Even though these approaches have less demanding requirements than the more comprehensive modeling approaches, they require non-trivial calibration using sufficient amounts of data, and they produce output signals that will not necessarily correspond to actual physiological sources in the brain. Furthermore, while ICA optimizes statistical independence of brain signals, and can thus lead to more compact signal representations, it does not guarantee to optimize the discriminability of different brain signals in different tasks. Consequently, these complex model-based and data-driven approaches may not be amenable or desirable for typical BCI experimentation. A more appropriate technique is Common Spatial Patterns (CSP) [29, 81]. This technique creates a spatial filter (i.e., weights for the signals acquired at different locations) that correspond to their importance in discriminating between different signal classes. Finally, even simpler deblurring filters have been shown to be effective and yet practical [56]. These filters are essentially spatial high-pass filters with fixed filtering characteristics. Typical realizations include Laplacian spatial filters and the Common Average Reference (CAR) filter.

A Laplacian spatial filter is comprised of discretized approximations of the second spatial derivative of the two-dimensional Gaussian distribution on the scalp surface, and attempts to invert the process that blurred the brain signals detected on the scalp [31]. The approximations are further simplified such that, at each time point t, the weighted sum of the potential s_i of the four nearest or next nearest electrodes is subtracted from the potential s_h at a center electrode for the small and large Laplacian, respectively (see Eq. 2.1, Fig. 2.12(a), and Fig. 2.12(b)).

$$s_h'(t) = s_h(t) - \sum_{i \in S_i} w_{h,i} s_i(t) \tag{2.1}$$

In this equation, the weight $w_{h,i}$ is a function of the distance $d_{h,i}$ between the electrode of interest h and its neighbor i:

$$w_{h,i} = \frac{\frac{1}{d_{h,i}}}{\sum_{i \in S_i} \frac{1}{d_{h,i}}} \tag{2.2}$$

In practice, this filter is often implemented simply by subtracting the average of the four next nearest neighbors (i.e., the weight for each neighbor is -0.25) from the center location.

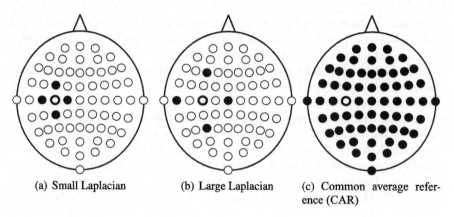

(a) Small Laplacian   (b) Large Laplacian   (c) Common average refer-
                                                             ence (CAR)

Fig. 2.12 Locations (*filled circles*) involved in different spatial filters applied to location C_3 (see Fig. 2.13) (*open circles*)

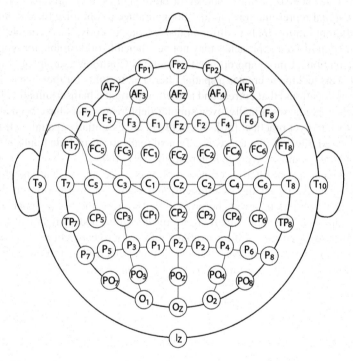

Fig. 2.13 Electrode designations for a 64-channel setup of the extended 10–20 system. (Redrawn from [94])

The Common Average Reference (CAR) filter, another possible spatial high-pass filter, is implemented by re-referencing the potential $s_h(t)$ of each electrode h at each time point t to an estimated reference that is calculated by averaging the signals from all H recorded electrodes (see Eq. 2.3 and Fig. 2.12(c)). In other words, a CAR

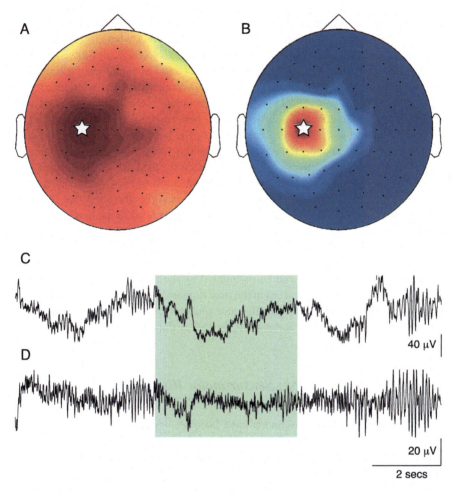

Fig. 2.14 Example of application of a Common Average Reference (CAR) spatial filter. Signals are spatially more specific and better highlight beta rhythm suppression during the period indicated by the *green bar* when a CAR filter is applied (**B** and **D**) compared to when it is not (**A** and **C**)

filter calculates the signal amplitude that is common to all electrodes ($\frac{1}{H} \sum_{i=1}^{H} s_i(t)$) and subtracts it from the signal $s_h(t)$ at each location. The CAR and Large Laplacian filters have been shown to provide comparable performance [56].

$$s'_h(t) = s_h(t) - \frac{1}{H} \sum_{i=1}^{H} s_i(t) \tag{2.3}$$

Whatever the realization of the spatial filter, its main purposes are to deblur the recorded signals so as to derive a more faithful representation of the sources within the brain, and/or to remove the influence of the reference electrode from the signal. The example in Fig. 2.14 illustrates results of this operation on signals recorded using EEG. In this example, the topographies illustrate the effect of a CAR filter on signals in space: in A, the unfiltered signal is spatially broad, whereas in B, the CAR-filtered signal emphasizes spatially local features. (Color indicates the correlation coefficient r calculated between the signal time course at each location (indicated by small dots) and the signal at the location indicated with the white star.) The signal time courses in C and D (which are recorded at the location marked with a star) illustrate the effect of a CAR filter on beta rhythms that are suppressed by right hand movement. The green bar indicates the period that the subject opened and closed her right hand. Beta rhythm oscillations (around 25 Hz) are suppressed during this period. This effect is more pronounced for the CAR-filtered signal in D compared to the unfiltered data in C. Also, application of the CAR filter removes the slow signal fluctuations seen in C. Thus, this example illustrates that the CAR filter removes some of the signal variance that is not related to the hand movement task.

The second and final step of feature extraction is the application of a temporal filter. Its purpose is to project the input signal in a representation (i.e., domain) in which the brain signals that can be modulated by the user are best expressed. The P300 potential is usually extracted in the time domain simply by averaging the brain responses to the different stimuli. Because mu/beta rhythms represent oscillatory activity, they are usually extracted in the frequency domain as described below.

2.4.3 Feature Extraction: Sensorimotor Rhythms

As described, imagined movements have been shown to produce changes in the mu (i.e., 8–12 Hz) or beta (i.e., 18–25 Hz) frequency band. Thus, for the processing of mu/beta rhythm (i.e., sensorimotor) signals, the frequency domain is most often used, although a recent paper suggested a matched filter in the time domain [38] that better captured the non-sinusoidal components of the mu rhythm. Several methods for the transformation of time domain into the frequency domain have been proposed, such as the Fast Fourier Transform (FFT), wavelet transform, and procedures based on autoregressive parameters. The requirements of BCI systems offer suggestions about which of these methods to select. BCIs are closed-loop systems that should provide feedback many times per second (e.g., typically, at more than 20 Hz) with a minimal group delay. In general, no method can concurrently achieve high resolution in time and frequency. However, the Maximum Entropy Method (MEM) [53], which is based on autoregressive modeling, has a higher temporal resolution (and thus reduced group delay) at a given frequency resolution compared to the FFT and wavelet transforms [53], and is thus advantageous in the context of BCI systems. Whatever its realization, the temporal filter used for mu/beta rhythms transforms time-domain signals $s'_e(t)$ into frequency-domain features $a_e(n)$.

As an example of its function, Fig. 2.15 illustrates application of the temporal filter to data collected using scalp-recorded EEG while a subject imagined hand movement or rested. CAR-filtered signals at each location were converted into the

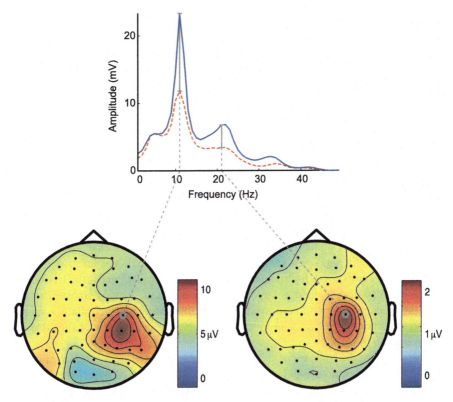

Fig. 2.15 Example analyses of the temporal filter. In this example, EEG was recorded during imagined left hand movement and during rest. The spectrum for rest (*blue solid line*) is different from that for imagined hand movement (*red dashed line*). This difference is sharply focused in frequency (see spectra *above*) and in space (see topographies *below*) and could be used for BCI control

frequency domain using the MEM algorithm and blocks of 400 ms length. The resulting spectra were averaged across blocks. This produced one average spectrum for each location and for each of the two conditions of imagined movement of the left hand and rest. (The example in Fig. 2.15 illustrates spectra at location C4.) The difference in the spectrum between imagined movement (red dashed line) and rest (blue solid line) is evident and is sharply focused in frequency (around 11 and 22 Hz) and space (over location C4 of right sensorimotor cortex). (While in this figure activity at 22 Hz may simply be a harmonic of activity at 11 Hz, evidence suggests that the relationship between mu and beta rhythms may be more complicated [57].) In other words, this particular subject can use particular motor imagery to change the signal amplitude at location C4 and 11/22 Hz and could thus communicate her intent using this particular type of imagery.

To facilitate understanding of the issues in BCI signal processing, the previous sections described the first component of BCI signal processing, feature extraction, which is composed of spatial filtering and temporal filtering. The purpose of feature extraction is to convert digitized brain signal samples that are recorded at various locations into features (e.g., frequency-domain spectra $a_e(n)$) that express the subject's intent. The second component of BCI signal processing, the translation algorithm, effects this intent by translating these features into device commands. The following section describes a common realization of the translation algorithm.

2.4.4 Translation Algorithm

The second step of BCI signal processing, the translation algorithm, is comprised mainly of a signal translation procedure that converts the set of brain signal features $a_e(n)$ into a set of output signals that control an output device. This translation has been traditionally accomplished using conventional classification/regression procedures. For example, studies have been using linear discriminant analysis [2], neural networks [34, 75], support vector machines [26, 30, 43, 65], and linear regression [54, 55]. To compensate for spontaneous changes in brain signals, the translation algorithm may also include a whitening procedure (e.g., a linear transformation) that produces signals with zero mean and a defined variance such that the output device does not have to account for changes in brain signal characteristics that are not related to the task.

We illustrate the function of a typical approach using an example realization of a translation algorithm using linear regression. Data were derived from the previous example shown in Fig. 2.15. Figure 2.16-A shows the distribution of data samples derived at C4 and 11/22 Hz (i.e., log transformed features $a_1(n)$, $a_2(n)$, respectively). Blue dots show samples that correspond to rest and red dots show samples that correspond to imagined left hand movement. Linear regression determined the coefficients of the linear function that minimized the error between the output of that function ($c(n)$) and arbitrary target values for the two classes (i.e., -1, $+1$). This procedure derived the coefficients of the linear function that could be used to translate the features into an output signal: $c(n) = 2{,}560a_1(n) + 4{,}582a_2(n)$. The histogram of the values of $c(n)$ calculated for the data from the rest class (blue) is different than that calculated for the data from the imagined hand class (red) (see Fig. 2.16-B), which indicates that the user had some level of control over this particular signal. To quantitatively evaluate this level of user control, we determined the value of r^2, i.e., the fraction of the variance in the output signal $c(n)$ that is associated with the classes. We then applied the same linear function to samples from all electrodes. Figure 2.16-C illustrates that, as expected, the signal difference is sharply focused over right sensorimotor cortex.

In summary, BCI signal processing is accomplished using two components. The first component, feature extraction, extracts brain signal features that reflect the user's intent. The second component, the translation algorithm, translates these signal features into output signals that can control an output device. This translation algorithm has traditionally been realized using a variety of classification/regression approaches. In summary, this chapter described relevant principles of BCI operation

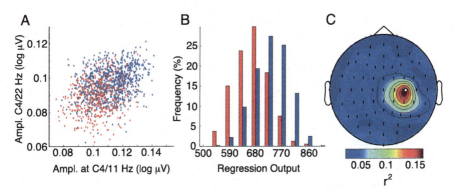

Fig. 2.16 Example of a typical translation algorithm that uses linear regression. **A**: Distribution of signal features (i.e., signal amplitudes at C4 and 11/22 Hz for rest (*blue dots*) and imagined left hand movement (*red dots*)). **B**: Histogram of output values calculated for each of the two classes using linear regression applied to the two features. **C**: As expected, the signal difference is focused over right sensorimotor cortex. See text for details

and associated techniques. The next chapter discusses the general concepts of the BCI2000 platform, and how BCI2000 can implement these techniques.

References

1. Allison, B.Z.: P3 or not P3: toward a better P300 BCI. PhD thesis, University of California, San Diego (2003)
2. Babiloni, F., Cincotti, F., Lazzarini, L., Millan, J., Mourino, J., Varsta, M., Heikkonen, J., Bianchi, L., Marciani, M.G.: Linear classification of low-resolution EEG patterns produced by imagined hand movements. IEEE Trans. Rehabil. Eng. **8**(2), 186–188 (2000)
3. Ball, T., Kern, M., Mutschler, I., Aertsen, A., Schulze-Bonhage, A.: Signal quality of simultaneously recorded invasive and non-invasive EEG. NeuroImage **46**(3), 708–716 (2009). doi:10.1016/j.neuroimage.2009.02.028. http://www.hubmed.org/display.cgi?uids= 19264143
4. Bashashati, A., Fatourechi, M., Ward, R.K., Birch, G.E.: A survey of signal processing algorithms in brain–computer interfaces based on electrical brain signals. J. Neural Eng. **4**(2), R32–R57 (2007). doi:10.1088/1741-2560/4/2/R03
5. Bayliss, J.D.: A flexible brain–computer interface. PhD thesis, University of Rochester, Rochester (2001). http://www.cs.rochester.edu/trs/robotics-trs.html
6. Birbaumer, N., Ghanayim, N., Hinterberger, T., Iversen, I., Kotchoubey, B., Kübler, A., Perelmouter, J., Taub, E., Flor, H.: A spelling device for the paralysed. Nature **398**(6725), 297–298 (1999)
7. Bullara, L.A., Agnew, W.F., Yuen, T.G., Jacques, S., Pudenz, R.H.: Evaluation of electrode array material for neural prostheses. Neurosurg. **5**(6), 681–686 (1979)
8. Chatrian, G.E.: The mu rhythm. In: Handbook of Electroencephalography and Clinical Neurophysiology. The EEG of the Waking Adult, pp. 46–69. Elsevier, Amsterdam (1976)
9. Chin, C.M., Popovic, M.R., Thrasher, A., Cameron, T., Lozano, A., Chen, R.: Identification of arm movements using correlation of electrocorticographic spectral components and kinematic recordings. J. Neural Eng. **4**(2), 146–158 (2007). doi:10.1088/1741-2560/4/2/014
10. Coyle, S., Ward, T., Markham, C., McDarby, G.: On the suitability of near-infrared (NIR) systems for next-generation brain–computer interfaces. Physiol. Meas. **25**(4), 815–822 (2004)

11. Coyle, S.M., Ward, T.E., Markham, C.M.: Brain–computer interface using a simplified functional near-infrared spectroscopy system. J. Neural Eng. **4**(3), 219–226 (2007). doi:10.1088/1741-2560/4/3/007

12. Crone, N.E., Miglioretti, D.L., Gordon, B., Sieracki, J.M., Wilson, M.T., Uematsu, S., Lesser, R.P.: Functional mapping of human sensorimotor cortex with electrocorticographic spectral analysis. i. Alpha and beta event-related desynchronization. Brain **121** (12), 2271–2299 (1998)

13. Crone, N.E., Miglioretti, D.L., Gordon, B., Lesser, R.P.: Functional mapping of human sensorimotor cortex with electrocorticographic spectral analysis. ii. Event-related synchronization in the gamma band. Brain **121** (12), 2301–2315 (1998)

14. Crone, N.E., Hao, L., Hart, J., Boatman, D., Lesser, R.P., Irizarry, R., Gordon, B.: Electrocorticographic gamma activity during word production in spoken and sign language. Neurol. **57**(11), 2045–2053 (2001)

15. Donchin, E.: Presidential address, 1980. Surprise!...Surprise? Psychophysiol. **18**(5), 493–513 (1981)

16. Donchin, E., Coles, M.: Is the P300 component a manifestation of context updating? Behav. Brain Sci. **11**(3), 357–427 (1988)

17. Donchin, E., Smith, D.B.: The contingent negative variation and the late positive wave of the average evoked potential. Electroencephalogr. Clin. Neurophysiol. **29**(2), 201–203 (1970)

18. Donchin, E., Heffley, E., Hillyard, S.A., Loveless, N., Maltzman, I., Ohman, A., Rösler, F., Ruchkin, D., Siddle, D.: Cognition and event-related potentials. ii. The orienting reflex and P300. Ann. N.Y. Acad. Sci. **425**, 39–57 (1984)

19. Donchin, E., Spencer, K.M., Wijesinghe, R.: The mental prosthesis: assessing the speed of a P300-based brain–computer interface. IEEE Trans. Rehabil. Eng. **8**(2), 174–179 (2000)

20. Donoghue, J., Nurmikko, A., Friehs, G., Black, M.: Development of neuromotor prostheses for humans. Suppl. Clin. Neurophysiol. **57**, 592–606 (2004)

21. Donoghue, J.P., Nurmikko, A., Black, M., Hochberg, L.R.: Assistive technology and robotic control using motor cortex ensemble-based neural interface systems in humans with tetraplegia. J. Physiol. **579**(3), 603–611 (2007). doi:10.1113/jphysiol.2006.127209

22. Farwell, L.A., Donchin, E.: Talking off the top of your head: toward a mental prosthesis utilizing event-related brain potentials. Electroencephalogr. Clin. Neurophysiol. **70**(6), 510–523 (1988)

23. Felton, E.A., Wilson, J.A., Williams, J.C., Garell, P.C.: Electrocorticographically controlled brain–computer interfaces using motor and sensory imagery in patients with temporary subdural electrode implants. Report of four cases. J. Neurosurg. **106**(3), 495–500 (2007)

24. Fisch, B.J.: Spehlmann's EEG Primer, 2nd edn. Elsevier, Amsterdam (1991)

25. Freeman, W.J., Holmes, M.D., Burke, B.C., Vanhatalo, S.: Spatial spectra of scalp EEG and EMG from awake humans. Clin. Neurophysiol. **114**, 1053–1068 (2003)

26. Garrett, D., Peterson, D.A., Anderson, C.W., Thaut, M.H.: Comparison of linear, nonlinear, and feature selection methods for EEG signal classification. IEEE Trans. Rehabil. Eng. **11**(2), 141–144 (2003)

27. Gastaut, H.: Etude electrocorticographique de la reactivite des rythmes rolandiques. Rev. Neurol. **87**, 176–182 (1952)

28. Graimann, B., Huggins, J.E., Schlögl, A., Levine, S.P., Pfurtscheller, G.: Detection of movement-related desynchronization patterns in ongoing single-channel electrocorticogram. IEEE Trans. Neural Syst. Rehabil. Eng. **11**(3), 276–281 (2003)

29. Guger, C., Ramoser, H., Pfurtscheller, G.: Real-time EEG analysis with subject-specific spatial patterns for a brain–computer interface (BCI). IEEE Trans. Rehabil. Eng. **8**(4), 447–456 (2000)

30. Gysels, E., Renevey, P., Celka, P.: SVM-based recursive feature elimination to compare phase synchronization computed from broadband and narrowband EEG signals in brain–computer interfaces. Signal Process. **85**(11), 2178–2189 (2005)

31. Hjorth, B.: Principles for transformation of scalp EEG from potential field into source distribution. J. Clin. Neurophysiol. **8**(4), 391–396 (1991)

32. Hochberg, L.R., Serruya, M.D., Friehs, G.M., Mukand, J.A., Saleh, M., Caplan, A.H., Branner, A., Chen, D., Penn, R.D., Donoghue, J.P.: Neuronal ensemble control of prosthetic devices by a human with tetraplegia. Nature **442**(7099), 164–171 (2006). doi:10.1038/nature04970

33. Hoffmann, U., Vesin, J.M., Ebrahimi, T., Diserens, K.: An efficient P300-based brain–computer interface for disabled subjects. J. Neurosci. Methods **167**(1), 115–125 (2008). doi:10.1016/j.jneumeth.2007.03.005

34. Huan, N.J., Palaniappan, R.: Neural network classification of autoregressive features from electroencephalogram signals for brain–computer interface design. J. Neural Eng. **1**(3), 142–150 (2004)

35. Jasper, H.H.: The ten twenty electrode system of the international federation. Electroencephalogr. Clin. Neurophysiol. **10**, 371–375 (1958)

36. Kostov, A., Polak, M.: Parallel man–machine training in development of EEG-based cursor control. IEEE Trans. Rehabil. Eng. **8**(2), 203–205 (2000)

37. Kozelka, J.W., Pedley, T.A.: Beta and mu rhythms. J. Clin. Neurophysiol. **7**, 191–207 (1990)

38. Krusienski, D.J., Schalk, G., McFarland, D.J., Wolpaw, J.R.: A mu-rhythm matched filter for continuous control of a brain–computer interface. IEEE Trans. Biomed. Eng. **54**(2), 273–280 (2007). doi:10.1109/TBME.2006.886661

39. Kübler, A., Kotchoubey, B., Hinterberger, T., Ghanayim, N., Perelmouter, J., Schauer, M., Fritsch, C., Taub, E., Birbaumer, N.: The Thought Translation Device: a neurophysiological approach to communication in total motor paralysis. Exp. Brain Res. **124**(2), 223–232 (1999)

40. Kübler, A., Nijboer, F., Mellinger, J., Vaughan, T.M., Pawelzik, H., Schalk, G., McFarland, D.J., Birbaumer, N., Wolpaw, J.R.: Patients with ALS can use sensorimotor rhythms to operate a brain–computer interface. Neurol. **64**(10), 1775–1777 (2005). doi:10.1212/01.WNL.0000158616.43002.6D

41. Lachaux, J.P., Fonlupt, P., Kahane, P., Minotti, L., Hoffmann, D., Bertrand, O., Baciu, M.: Relationship between task-related gamma oscillations and bold signal: new insights from combined fMRI and intracranial EEG. Hum. Brain Mapp. **28**(12), 1368–1375 (2007). doi:10.1002/hbm.20352

42. LaConte, S.M., Peltier, S.J., Hu, X.P.: Real-time fMRI using brain-state classification. Hum. Brain Mapp. **28**(10), 1033–1044 (2007). doi:10.1002/hbm.20326. http://www.hubmed.org/display.cgi?uids=17133383

43. Lal, T.N., Schroder, M., Hinterberger, T., Weston, J., Bogdan, M., Birbaumer, N., Schölkopf, B.: Support vector channel selection in BCI. IEEE Trans. Biomed. Eng. **51**(6), 1003–1010 (2004)

44. Le, J., Gevins, A.: Method to reduce blur distortion from EEG's using a realistic head model. IEEE Trans. Biomed. Eng. **40**(6), 517–528 (1993)

45. Lebedev, M.A., Carmena, J.M., O'Doherty, J.E., Zacksenhouse, M., Henriquez, C.S., Principe, J.C., Nicolelis, M.A.: Cortical ensemble adaptation to represent velocity of an artificial actuator controlled by a brain–machine interface. J. Neurosci. **25**(19), 4681–4693 (2005). doi:10.1523/JNEUROSCI.4088-04.2005

46. Leuthardt, E., Schalk, G., JR, J.W., Ojemann, J., Moran, D.: A brain–computer interface using electrocorticographic signals in humans. J. Neural Eng. **1**(2), 63–71 (2004)

47. Leuthardt, E., Miller, K., Schalk, G., Rao, R., Ojemann, J.: Electrocorticography-based brain computer interface – the Seattle experience. IEEE Trans. Neural Syst. Rehabil. Eng. **14**, 194–198 (2006)

48. Leuthardt, E., Miller, K., Anderson, N., Schalk, G., Dowling, J., Miller, J., Moran, D., Ojemann, J.: Electrocorticographic frequency alteration mapping: a clinical technique for mapping the motor cortex. Neurosurg. **60**, 260–270, discussion 270–271 (2007). doi:10.1227/01.NEU.0000255413.70807.6E

49. Loeb, G.E., Walker, A.E., Uematsu, S., Konigsmark, B.W.: Histological reaction to various conductive and dielectric films chronically implanted in the subdural space. J. Biomed. Mater. Res. **11**(2), 195–210 (1977). doi:10.1002/jbm.820110206

50. Lotte, F., Congedo, M., Lécuyer, A., Lamarche, F., Arnaldi, B.: A review of classification algorithms for EEG-based brain–computer interfaces. J. Neural Eng. **4**(2), 1–1 (2007). doi:10.1088/1741-2560/4/2/R01

51. Makeig, S., Jung, T., Bell, A., Sejnowski, T.: Independent component analysis of electroencephalographic data. In: Advances in Neural Information Processing Systems, vol. 8, pp. 145–151. MIT Press, Cambridge (1996)

52. Margalit, E., Weiland, J., Clatterbuck, R., Fujii, G., Maia, M., Tameesh, M., Torres, G., D'Anna, S., Desai, S., Piyathaisere, D., Olivi, A., de Juan, E.J., Humayun, M.: Visual and electrical evoked response recorded from subdural electrodes implanted above the visual cortex in normal dogs under two methods of anesthesia. J. Neurosci. Methods **123**(2), 129–137 (2003)

53. Marple, S.L.: Digital Spectral Analysis: With Applications. Prentice–Hall, Englewood Cliffs (1987)

54. McFarland, D.J., Neat, G.W., Wolpaw, J.R.: An EEG-based method for graded cursor control. Psychobiol. **21**, 77–81 (1993)

55. McFarland, D.J., Lefkowicz, T., Wolpaw, J.R.: Design and operation of an EEG-based brain–computer interface (BCI) with digital signal processing technology. Behav. Res. Methods Instrum. Comput. **29**, 337–345 (1997)

56. McFarland, D.J., McCane, L.M., David, S.V., Wolpaw, J.R.: Spatial filter selection for EEG-based communication. Electroencephalogr. Clin. Neurophysiol. **103**(3), 386–394 (1997)

57. McFarland, D.J., Miner, L.A., Vaughan, T.M., Wolpaw, J.R.: Mu and beta rhythm topographies during motor imagery and actual movements. Brain Topogr. **12**(3), 177–186 (2000)

58. McFarland, D., Anderson, C.W., Müller, K.R., Schlögl, A., Krusienski, D.J.: BCI meeting 2005 – workshop on BCI signal processing: feature extraction and translation. IEEE Trans. Neural Syst. Rehabil. Eng. **14**(2), 135–138 (2006)

59. McFarland, D.J., Krusienski, D.J., Sarnacki, W.A., Wolpaw, J.R.: Emulation of computer mouse control with a noninvasive brain–computer interface. J. Neural Eng. **5**(2), 101–110 (2008). doi:10.1088/1741-2560/5/2/001. http://www.hubmed.org/display.cgi?uids=18367779

60. Mellinger, J., Schalk, G., Braun, C., Preissl, H., Rosenstiel, W., Birbaumer, N., Kübler, A.: An MEG-based brain–computer interface (BCI). NeuroImage **36**(3), 581–593 (2007). doi:10.1016/j.neuroimage.2007.03.019

61. Millán, J. del R., Renkens, F., Mouriño, J., Gerstner, W.: Noninvasive brain-actuated control of a mobile robot by human EEG. IEEE Trans. Biomed. Eng. **51**(6), 1026–1033 (2004)

62. Miller, K., Leuthardt, E., Schalk, G., Rao, R., Anderson, N., Moran, D., Miller, J., Ojemann, J.: Spectral changes in cortical surface potentials during motor movement. J. Neurosci. **27**, 2424–2432 (2007). doi:10.1523/JNEUROSCI.3886-06.2007. http://www.jneurosci.org/cgi/content/abstract/27/9/2424

63. Morgan, S.T., Hansen, J.C., Hillyard, S.A.: Selective attention to stimulus location modulates the steady-state visual evoked potential. Proc. Natl. Acad. Sci. USA **93**(10), 4770–4774 (1996)

64. Müller, K., Blankertz, B.: Toward noninvasive brain–computer interfaces. IEEE Signal Process. Mag. **23**(5), 126–128 (2006)

65. Müller, K.R., Anderson, C.W., Birch, G.E.: Linear and nonlinear methods for brain–computer interfaces. IEEE Trans. Rehabil. Eng. **11**(2), 165–169 (2003)

66. Müller, K.R., Tangermann, M., Dornhege, G., Krauledat, M., Curio, G., Blankertz, B.: Machine learning for real-time single-trial EEG-analysis: from brain–computer interfacing to mental state monitoring. J. Neurosci. Methods **167**(1), 82–90 (2008). doi:10.1016/j.jneumeth.2007.09.022. http://www.hubmed.org/display.cgi?uids=18031824

67. Musallam, S., Corneil, B.D., Greger, B., Scherberger, H., Andersen, R.A.: Cognitive control signals for neural prosthetics. Science **305**(5681), 258–262 (2004). doi:10.1126/science.1097938

68. Neshige, R., Murayama, N., Tanoue, K., Kurokawa, H., Igasaki, T.: Optimal methods of stimulus presentation and frequency analysis in P300-based brain–computer interfaces for patients with severe motor impairment. Suppl. Clin. Neurophysiol. **59**, 35–42 (2006)

69. Niedermeyer, E.: The normal EEG of the waking adult. In: Niedermeyer, E., Lopes da Silva, F.H. (eds.) Electroencephalography: Basic Principles, Clinical Applications and Related Fields, 4th edn., pp. 149–173. Williams and Wilkins, Baltimore (1999)

70. Pfurtscheller, G.: EEG event-related desynchronization (ERD) and event-related synchronization (ERS). In: Niedermeyer, E., Lopes da Silva, F.H. (eds.) Electroencephalography: Basic Principles, Clinical Applications and Related Fields, 4th edn., pp. 958–967. Williams and Wilkins, Baltimore (1999)

71. Pfurtscheller, G., Berghold, A.: Patterns of cortical activation during planning of voluntary movement. Electroencephalogr. Clin. Neurophysiol. **72**, 250–258 (1989)

72. Pfurtscheller, G., Neuper, C.: Motor imagery activates primary sensorimotor area in humans. Neurosci. Lett. **239**, 65–68 (1997)

73. Pfurtscheller, G., Flotzinger, D., Kalcher, J.: Brain–computer interface – a new communication device for handicapped persons. J. Microcomput. Appl. **16**, 293–299 (1993)

74. Pfurtscheller, G., Neuper, C., Kalcher, J.: 40-Hz oscillations during motor behavior in man. Neurosci. Lett. **164**(1–2), 179–182 (1993)

75. Pfurtscheller, G., Neuper, C., Flotzinger, D., Pregenzer, M.: EEG-based discrimination between imagination of right and left hand movement. Electroencephalogr. Clin. Neurophysiol. **103**(6), 642–651 (1997)

76. Pfurtscheller, G., Guger, C., Müller, G., Krausz, G., Neuper, C.: Brain oscillations control hand orthosis in a tetraplegic. Neurosci. Lett. **292**(3), 211–214 (2000)

77. Pfurtscheller, G., Graimann, B., Huggins, J.E., Levine, S.P., Schuh, L.A.: Spatiotemporal patterns of beta desynchronization and gamma synchronization in corticographic data during self-paced movement. Clin. Neurophysiol. **114**(7), 1226–1236 (2003)

78. Piccione, F., Giorgi, F., Tonin, P., Priftis, K., Giove, S., Silvoni, S., Palmas, G., Beverina, F.: P300-based brain computer interface: reliability and performance in healthy and paralysed participants. Clin. Neurophysiol. **117**(3), 531–537 (2006). doi:10.1016/j.clinph.2005.07.024

79. Pistohl, T., Ball, T., Schulze-Bonhage, A., Aertsen, A., Mehring, C.: Prediction of arm movement trajectories from ECoG-recordings in humans. J. Neurosci. Methods **167**(1), 105–114 (2008)

80. Pritchard, W.S.: Psychophysiology of P300. Psychol. Bull. **89**(3), 506–540 (1981)

81. Ramoser, H., Müller-Gerking, J., Pfurtscheller, G.: Optimal spatial filtering of single trial EEG during imagined hand movement. IEEE Trans. Rehabil. Eng. **8**(4), 441–446 (2000)

82. Ramsey, N.F., van de Heuvel, M.P., Kho, K.H., Leijten, F.S.: Towards human BCI applications based on cognitive brain systems: an investigation of neural signals recorded from the dorsolateral prefrontal cortex. IEEE Trans. Neural Syst. Rehabil. Eng. **14**(2), 214–217 (2006). http://www.hubmed.org/display.cgi?uids=16792297

83. Sanchez, J.C., Gunduz, A., Carney, P.R., Principe, J.C.: Extraction and localization of mesoscopic motor control signals for human ECoG neuroprosthetics. J. Neurosci. Methods **167**(1), 63–81 (2008). doi:10.1016/j.jneumeth.2007.04.019

84. Santhanam, G., Ryu, S.I., Yu, B.M., Afshar, A., Shenoy, K.V.: A high-performance brain–computer interface. Nature **442**(7099), 195–198 (2006). doi:10.1038/nature04968

85. Schalk, G., McFarland, D., Hinterberger, T., Birbaumer, N., Wolpaw, J.: BCI2000: a general-purpose brain–computer interface (BCI) system. IEEE Trans. Biomed. Eng. **51**, 1034–1043 (2004)

86. Schalk, G., Kubánek, J., Miller, K.J., Anderson, N.R., Leuthardt, E.C., Ojemann, J.G., Limbrick, D., Moran, D., Gerhardt, L.A., Wolpaw, J.R.: Decoding two-dimensional movement trajectories using electrocorticographic signals in humans. J. Neural Eng. **4**(3), 264–275 (2007). doi:10.1088/1741-2560/4/3/012

87. Schalk, G., Miller, K.J., Anderson, N.R., Wilson, J.A., Smyth, M.D., Ojemann, J.G., Moran, D.W., Wolpaw, J.R., Leuthardt, E.C.: Two-dimensional movement control using electrocorticographic signals in humans. J. Neural Eng. **5**(1), 75–84 (2008). doi:10.1088/1741-2560/5/1/008

88. Sellers, E.W., Donchin, E.: A P300-based brain–computer interface: initial tests by ALS patients. Clin. Neurophysiol. **117**(3), 538–548 (2006). doi:10.1016/j.clinph.2005.06.027

89. Sellers, E.W., Kübler, A., Donchin, E.: Brain–computer interface research at the University of South Florida Cognitive Psychophysiology Laboratory: the P300 Speller. IEEE Trans. Neural Syst. Rehabil. Eng. **14**(2), 221–224 (2006)

90. Sellers, E.W., Krusienski, D.J., McFarland, D.J., Vaughan, T.M., Wolpaw, J.R.: A P300 event-related potential brain–computer interface (BCI): the effects of matrix size and inter stimulus interval on performance. Biol. Psychol. **73**(3), 242–252 (2006). doi:10.1016/j.biopsycho.2006.04.007

91. Serby, H., Yom-Tov, E., Inbar, G.F.: An improved P300-based brain–computer interface. IEEE Trans. Neural Syst. Rehabil. Eng. **13**(1), 89–98 (2005)

92. Serruya, M., Hatsopoulos, N., Paninski, L., Fellows, M., Donoghue, J.: Instant neural control of a movement signal. Nature **416**(6877), 141–142 (2002)

93. Shain, W., Spataro, L., Dilgen, J., Haverstick, K., Retterer, S., Isaacson, M., Saltzman, M., Turner, J.: Controlling cellular reactive responses around neural prosthetic devices using peripheral and local intervention strategies. IEEE Trans. Neural Syst. Rehabil. Eng. **11**, 186–188 (2003)

94. Sharbrough, F., Chatrian, G., Lesser, R., Luders, H., Nuwer, M., Picton, T.: American electroencephalographic society guidelines for standard electrode position nomenclature. Electroencephalogr. Clin. Neurophysiol. **8**, 200–202 (1991)

95. Shenoy, K., Meeker, D., Cao, S., Kureshi, S., Pesaran, B., Buneo, C., Batista, A., Mitra, P., Burdick, J., Andersen, R.: Neural prosthetic control signals from plan activity. Neurorep. **14**(4), 591–596 (2003)

96. Sinai, A., Bowers, C.W., Crainiceanu, C.M., Boatman, D., Gordon, B., Lesser, R.P., Lenz, F.A., Crone, N.E.: Electrocorticographic high gamma activity versus electrical cortical stimulation mapping of naming. Brain **128**(7), 1556–1570 (2005). doi:10.1093/brain/awh491

97. Sitaram, R., Caria, A., Birbaumer, N.: Hemodynamic brain–computer interfaces for communication and rehabilitation. Neural Netw. **22**(9), 1320–1328 (2009). doi:10.1016/j.neunet.2009.05.009. http://www.hubmed.org/display.cgi?uids=19524399

98. Sitaram, R., Caria, A. Veit, R., Gaber, T., Rota, G., Kübler, A., Birbaumer, N.: fMRI brain–computer interface: a tool for neuroscientific research and treatment. Comput. Intell. Neurosci. **2007**, Article ID 25487 (10 pages) (2007). doi:10.1155/2007/25487

99. Staba, R.J., Wilson, C.L., Bragin, A., Fried, I., Engel, J.: Quantitative analysis of high-frequency oscillations (80–500 Hz) recorded in human epileptic hippocampus and entorhinal cortex. J. Neurophysiol. **88**(4), 1743–1752 (2002)

100. Stice, P., Muthuswamy, J.: Assessment of gliosis around moveable implants in the brain. J. Neural Eng. **6**(4), 046004 (2009). doi:10.1088/1741-2560/6/4/046004

101. Sutter, E.E.: The brain response interface: communication through visually guided electrical brain responses. J. Microcomput. Appl. **15**, 31–45 (1992)

102. Sutton, S., Braren, M., Zubin, J., John, E.R.: Evoked-potential correlates of stimulus uncertainty. Science **150**(700), 1187–1188 (1965)

103. Taylor, D.M., Tillery, S.I., Schwartz, A.B.: Direct cortical control of 3D neuroprosthetic devices. Science **296**, 1829–1832 (2002)

104. Toro, C., Cox, C., Friehs, G., Ojakangas, C., Maxwell, R., Gates, J.R., Gumnit, R.J., Ebner, T.J.: 8–12 Hz rhythmic oscillations in human motor cortex during two-dimensional arm movements: evidence for representation of kinematic parameters. Electroencephalogr. Clin. Neurophysiol. **93**(5), 390–403 (1994)

105. Turner, J.N., Ancin, H., Becker, D., Szarowski, D.H., Holmes, M., O'Connor, N., Wang, M., Holmes, T.J., Roysam, B.: Automated image analysis technologies for biological 3-d light microscopy. Int. J. Imaging Syst. Technol., Spec. Issue Microsc. **8**, 240–254 (1997)

106. Vaughan, T.M., McFarland, D.J., Schalk, G., Sarnacki, W.A., Krusienski, D.J., Sellers, E.W., Wolpaw, J.R.: The Wadsworth BCI research and development program: at home with BCI. IEEE Trans. Neural Syst. Rehabil. Eng. **14**(2), 229–233 (2006)

107. Walter, W.G., Cooper, R., Aldridge, V.J., McCallum, W.C., Winter, A.L.: Contingent negative variation: an electric sign of sensorimotor association and expectancy in the human brain. Nature **203**, 380–384 (1964)

108. Weiskopf, N., Veit, R., Erb, M., Mathiak, K., Grodd, W., Goebel, R., Birbaumer, N.: Physiological self-regulation of regional brain activity using real-time functional magnetic resonance imaging (fMRI): methodology and exemplary data. NeuroImage **19**(3), 577–586 (2003)

109. Weiskopf, N., Mathiak, K., Bock, S.W., Scharnowski, F., Veit, R., Grodd, W., Goebel, R., Birbaumer, N.: Principles of a brain–computer interface (BCI) based on real-time functional magnetic resonance imaging (fMRI). IEEE Trans. Biomed. Eng. **51**(6), 966–970 (2004)

110. Weiskopf, N., Scharnowski, F., Veit, R., Goebel, R., Birbaumer, N., Mathiak, K.: Self-regulation of local brain activity using real-time functional magnetic resonance imaging (fMRI). J. Physiol. Paris **98**(4–6), 357–373 (2004). doi:10.1016/j.jphysparis.2005.09.019

111. Wilson, J., Felton, E., Garell, P., Schalk, G., Williams, J.: ECoG factors underlying multimodal control of a brain–computer interface. IEEE Trans. Neural Syst. Rehabil. Eng. **14**, 246–250 (2006)

112. Wolpaw, J., Birbaumer, N.: Brain–computer interfaces for communication and control. In: Selzer, M., Clarke, S., Cohen, L., Duncan, P., Gage, F. (eds.) Textbook of Neural Repair and Rehabilitation; Neural Repair and Plasticity, pp. 602–614. Cambridge University Press, Cambridge (2006)

113. Wolpaw, J.R., McFarland, D.J.: Multichannel EEG-based brain–computer communication. Electroencephalogr. Clin. Neurophysiol. **90**(6), 444–449 (1994)

114. Wolpaw, J.R., McFarland, D.J.: Control of a two-dimensional movement signal by a noninvasive brain–computer interface in humans. Proc. Natl. Acad. Sci. USA **101**(51), 17849–17854 (2004). doi:10.1073/pnas.0403504101. http://www.hubmed.org/display.cgi?uids=15585584

115. Wolpaw, J., McFarland, D., Cacace, A.: Preliminary studies for a direct brain-to-computer parallel interface. In: Projects for Persons with Disabilities. IBM Technical Symposium, pp. 11–20 (1986)

116. Wolpaw, J.R., McFarland, D.J., Neat, G.W., Forneris, C.A.: An EEG-based brain–computer interface for cursor control. Electroencephalogr. Clin. Neurophysiol. **78**(3), 252–259 (1991)

117. Wolpaw, J.R., Birbaumer, N., McFarland, D.J., Pfurtscheller, G., Vaughan, T.M.: Brain–computer interfaces for communication and control. Electroencephalogr. Clin. Neurophysiol. **113**(6), 767–791 (2002)

118. Yoo, S.S., Fairneny, T., Chen, N.K., Choo, S.E., Panych, L.P., Park, H., Lee, S.Y., Jolesz, F.A.: Brain–computer interface using fMRI: spatial navigation by thoughts. Neurorep. **15**(10), 1591–1595 (2004)

119. Yuen, T.G., Agnew, W.F., Bullara, L.A.: Tissue response to potential neuroprosthetic materials implanted subdurally. Biomaterials **8**(2), 138–141 (1987)

Chapter 3
Introducing BCI2000

3.1 Motivation

As described in the previous chapter, many studies over the past two decades have shown that non-muscular communication and control is possible and might serve useful purposes for those who cannot use conventional technologies. At the same time, the performance of this new technology is still modest. The potential value of this new technology will depend largely on the degree to which its performance, i.e., information transfer rate between the user and the BCI system, can be increased.

Many factors determine the performance of a BCI system. These factors include the brain signals measured, the signal processing methods that extract signal features, the algorithms that translate these features into device commands, the output devices that execute these commands, the feedback provided to the user, and the characteristics of the user. Thus, future progress requires systematic well-controlled studies that evaluate and compare alternative signals and combinations of signals, alternative feature extraction methods and translation algorithms, and alternative communication and control applications in different user populations. In consequence, a typical research and development program focused on human BCI research will usually run several studies at the same time, possibly even in different locations, and most often by different personnel.

These requirements imply the need for a software tool that facilitates the implementation of any BCI system, and that facilitates the collaboration of multiple laboratories on algorithm design, experimental design, or data analysis. In other words, standardized software or procedural mechanisms for development of BCI methods and their components, for data exchange, and for the appropriate documentation of relevant configuration parameters are necessary. Furthermore, implementation of experimental paradigms in other laboratories should be as easy as possible. Unfortunately, the typical BCI system does not readily support such systematic research and development. While a few systems have attempted to address some aspects of these issues (e.g., [1, 2, 5, 9]), BCI research has usually consisted of a technical demonstration of a very specific BCI paradigm, i.e., of a demonstration that a certain brain signal recorded and measured in a certain way, and translated into control

G. Schalk, J. Mellinger, *A Practical Guide to Brain–Computer Interfacing with BCI2000*,
© Springer-Verlag London Limited 2010

commands by a certain algorithm, can control a certain device for one or a few users [15]. Importantly, this situation has not changed by the growing maturization of tools that support rapid prototyping, such as Matlab or LabView. While these tools make it easier to implement one particular BCI paradigm that is run by one person with one particular set of hardware components, such prototyped solutions are typically not robust and/or general enough for use by other people in other environments. In sum, BCI demonstrations have often been based on prototypes that do not support the requirements set forth above.

In recognition of this situation, we set out to develop, test, and disseminate a general-purpose BCI research and development system, called BCI2000, that can facilitate such systematic studies. The goals of the BCI2000 project are (1) to create a system that can facilitate the implementation and collaborative use of any BCI system; (2) to incorporate into this system support for the most commonly used BCI methods; and (3) to disseminate the system and associated documentation to other laboratories. BCI2000 should thus facilitate progress in laboratory and clinical BCI research by reducing the time, effort, and expense of testing new BCI methods, by providing a standardized data format for offline analyses, and by allowing groups lacking high-level software expertise to engage in BCI research.

Over the past nine years, the BCI2000 platform has evolved into a robust software system that can support a wide array of BCI and related studies. To date, BCI2000 has been used to implement BCI methods that can use more than fifteen data acquisition devices, that can make use of sensorimotor rhythms, cortical surface rhythms, slow cortical potentials, and the P300 potential, and that can provide the outputs needed for different kinds of output devices (such as computer cursors, robotic arms, sequential menus, etc.). The continuing grant support for the BCI2000 project also ensures that the system continues to evolve and improve. To facilitate integration in other environments, the BCI2000 software can run on standard PC hardware, and supports a variety of data acquisition devices. Because it is written in C++, it makes efficient use of computational resources and can satisfy the real-time requirements of BCI operation.

3.2 The Design of the BCI2000 Platform

3.2.1 A Common Model

BCI2000 is based on a model that can describe any BCI system and that is similar to the one described in [7]. This model (see Fig. 3.1 for a simplified diagram), consists of four modules that communicate with each other: Source (Data Acquisition and Storage), Signal Processing, User Application, and Operator Interface. The modules are separate programs that communicate through a TCP/IP-based protocol. This protocol can transmit all information (e.g., signals or variables) needed for operation. Thus, the protocol does not need to be changed when changes are made in a module. Brain signals are processed synchronously, in blocks containing a fixed number of

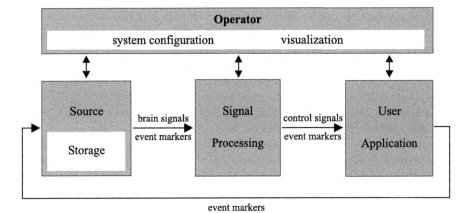

Fig. 3.1 BCI2000 design. BCI2000 consists of four modules: Operator, Source, Signal Processing, and Application. The Operator module acts as a central relay for system configuration and online presentation of results to the investigator. It also defines onset and offset of operation. During operation, information (i.e., signals, parameters, or event markers) is communicated from Source to Signal Processing to User Application and back to Source

samples that are acquired by the Source module. We chose synchronous processing over asynchronous processing, because it makes it more practical to guarantee system performance, and it allows for a very generic mechanism to relate the timing of stimulus presentation to the timing of data acquisition. During system operation, each time a new block of data is acquired, the Source module sends it to Signal Processing, which extracts signal features, translates those features into control signals, and sends them on to the Application module. Finally, the Application module sends the resulting event markers back to the Source module where they and the raw signals are stored to disk. The contents of the data file thus allow for full reconstruction of an experimental session during offline analyses.

The choice of block size is determined by processing resources as well as timing precision considerations. In the BCI2000 on-line system, the duration of a block corresponds to the temporal resolution of stimulus presentation, e.g., cursor update rate. This suggests that a small block size is desirable. On the other hand, real-time operation implies that the average time required for processing a block and communicating it between modules (roundtrip time) be less than a block's duration. Thus, processing resources (and network latencies in a distributed system) impose a lower limit on the block size. A typical configuration, i.e., sampling 16–64 channels at 256 Hz and processing blocks of eight samples each, will result in a feedback update rate of 32 Hz with small output latency. When using data sources that can only send data at specific block sizes, BCI2000 block size will be further constrained to multiples of the source's block size. Because the latency between completed acquisition of a data block and stimulus presentation is very small, this synchronous relationship of data acquisition and stimulus presentation provides a practical and generic way to associate event markers with brain signal samples.

To maximize interchangeability and independence in BCI2000, we based system design on principles applicable to all BCIs and implemented these principles using current techniques of object-oriented software design.

The TCP/IP-based communication protocol can transmit all information (e.g., signals or variables) needed for operation. Thus, it does not need to be changed when changes are made in a module. The information that passes from one module to another is highly standardized to minimize the dependencies between modules.[1] Each necessary BCI function is placed in the module to which it logically belongs. For example, because each processing cycle is initiated by the acquisition of a block of data samples, Source acts as BCI2000's system clock. Similarly, because feedback control varies with the application (e.g., fixed-length vs. user-paced applications), it is placed in User Application. This principle further reduces interdependence between modules.

The four modules and their communication protocol do not place constraints on the number of signal channels or their sampling rate, the number of system parameters or event markers, the complexity of signal processing, the timing of operation, or the number of signals that control the output device. Thus, these factors are limited only by the capacities of the hardware used.

3.2.2 Source Module and File Format

The Source module acquires brain signals and passes calibrated signal samples on to the Signal Processing module. The Source module consists of a data acquisition component, and a data storage component that implements the native BCI2000 file format, as well as EDF, which is a data format popular in sleep research [6], and GDF, a variant of EDF designed for BCI applications [11]. The data acquisition component has a number of implementations. The implementations for the g.USBamp and g.MOBIlab devices from g.tec are directly supported by the BCI2000 development team, i.e., they are part of the BCI2000 *core* distribution. A number of implementations for hardware from other manufacturers (i.e., BioSemi, BrainProducts, Cleveland Medical Devices, Data Translation, Electrical Geodesics, Measurement Computing, National Instruments, Neuroscan, OpenEEG, Tucker-Davis Technologies) have been provided by the BCI2000 user community.

The BCI2000 file format consists of an ASCII header that defines all parameters used for this particular experimental session, followed by binary signal sample and event marker values. The file format can accommodate any number of signal

---

[1]User applications might employ distinctive stimuli to evoke brain responses (such as the P300 speller included with BCI2000). Because these BCI paradigms depend on the brain's responses to the stimuli, their User Applications cannot be interchanged with ones that do not provide the stimuli. In other situations, online adaptations occurring in the Signal Processing module depend on the feedback provided to the user. Thus, a certain degree of module interdependence is sometimes unavoidable.

channels, system parameters, or event markers, and thus can accommodate any experimental protocol. It supports 16- and 32-bit integer formats as well as a 32-bit floating point format.

3.2.3 Signal Processing Module

The Signal Processing module converts signals from the brain into signals that control an output device. In the current BCI2000 implementations, this conversion is done in two stages, feature extraction and translation, and realized using a chain of signal filters, each of which transforms an input signal into an output signal. The individual signal filters are designed to be independent of each other and can thus be combined or interchanged without affecting others.

The first stage, feature extraction, is currently comprised of two filters. The first filter can implement any linear spatial filtering operation by calculating a matrix multiplication of the input signals with a spatial filtering matrix. The second filter is called "temporal filter." The current BCI2000 core distribution comes with three different variations of this temporal filter: autoregressive spectral estimation; spectral estimation based on the Fast Fourier Transform (FFT); and a filter that averages evoked responses (e.g., P300s). The second stage, feature translation, translates the extracted signal features into device-independent control signals. This is done by two filters. The first applies a linear classifier, and the second filter normalizes the output signals such that they have zero mean and a specific value range. The output of this procedure is the output of the Signal Processing module.

3.2.4 User Application Module

The User Application module receives control signals from Signal Processing and uses them to drive an application. In most present-day BCIs, the user application is presented visually on a computer screen and consists of the selection of targets, letters, or icons.

Existing User Application modules in BCI2000 implement very capable versions of popular feedback paradigms: a three-dimensional cursor movement paradigm (Cursor Task); a matrix spelling application based on P300 evoked potentials (P3 Speller); and presentation of auditory and visual stimuli with optional feedback of evoked potential classification results (Stimulus Presentation). Figure 3.2-A–C shows the display to the user for these three applications, respectively.

3.2.5 Operator Module

The Operator module provides the investigator with a graphical interface that displays current system parameters and real-time analysis results (e.g., frequency

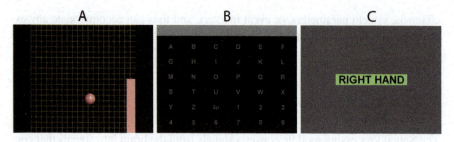

Fig. 3.2 Displays for the three BCI2000 User Applications. **A**: Cursor movement to a variable number of targets (i.e., Cursor Task). **B**: A spelling application based on P300 evoked potentials (i.e., P3 Speller). **C**: Auditory/visual stimulation program (i.e., Stimulus Presentation). In **A**, the cursor moves from a programmable location towards one of N programmable targets. In **B**, rows and columns of the matrix flash in a block-randomized fashion. In **C**, a series of programmable stimuli are presented in sequence

spectra) communicated to it from other modules. It allows the investigator to start, stop, suspend, resume, or reconfigure system operation. In a typical BCI2000 configuration, user feedback is displayed on one monitor, and the Operator module's graphical interface (i.e., the interface to the investigator) is displayed on a second monitor.

3.2.6 System Variables

BCI2000 incorporates three types of system variables: parameters, event markers, and signals. System parameters are those variables that do not change throughout a data file (i.e., during a specified period of online operation). In contrast, event markers record events that occur during operation and that can change from one data sample to the next. The inclusion of all event markers in the data file allows full offline reconstruction and analysis of the experimental session. Each module has access to these event markers, and can modify and/or simply monitor them. Finally, system signals are functions of the user's brain signals that are received and modified by the modules.

Each module can request that the Operator module create any number of system parameters (of different data types such as numbers, vectors, matrices, or strings) or event markers (each 1–16 bits long). For example, the Source module might request a parameter that defines the signal's sampling rate. This parameter is constant during some defined period of online operation, is available to all other modules, and is automatically recorded in the data file. Similarly, a Signal Processing filter designed to detect artifacts (such as those created by muscle movements) might request an event marker with which to mark artifacts in the signal, and an Application module might request an event marker to record stimulus conditions. These different system variables are automatically communicated and supported throughout the system, are aligned with signal samples, and stored.

3.3 BCI2000 Advantages

The generic concepts described above are of particular advantage for large research and development programs. For example, because of the standardized protocol between modules, different realizations of the different modules can usually be mixed with other modules. For example, different Source modules (providing support for different data acquisition devices) can be utilized with the same signal processing routines and feedback modalities. This greatly facilitates the development of the complete experiments, or parts of experiments, in different locations and environments.

Because the graphical interface to the investigator in the Operator program is generated dynamically based on the parameterization requirements of the particular data acquisition device, signal processing routines, and experimental protocol, the same Operator program (and not one for each possible combination of these routines) can be used in different environments or for different experiments.

Because the data format adapts to the particular set of event markers requested by the different modules, introduction of a new event marker (e.g., to report an artifact detected by a new signal processing routine) only involves a small change in the component that realizes the new signal processing component. The data format is automatically adapted by the system and all BCI2000 components interacting with the data format (e.g., file viewer, data export into Matlab) are able to work with these new types of data without any change.

In summary, the BCI2000 platform implements a number of general concepts that are applicable to a wide array of different experiments. Thus, it provides optimum benefits to interactive and collaborative situations that involve a number of experiments, locations, and personnel.

3.4 System Requirements and Real-Time Processing

As a run-time environment, the current implementation of BCI2000 requires Microsoft Windows 2000 or a more recent Windows-based operating system, any recent desktop or laptop computer, and one of the data acquisition devices that BCI2000 supports.

A BCI system must acquire and process brain signals (potentially from many channels at high sampling rates) and respond with appropriate output within a short time period (i.e., latency) with minimal variation (i.e., latency jitter). To give an impression of the timing performance of BCI2000 in actual online operation, we evaluated system performance in two different signal processing/task configurations. The hardware platform used for this test was a Mac Pro workstation with two 2.8 GHz quad-core Intel Xeon processors, 6 GB RAM, NVIDIA 8800GT (512 MB) video card, running Windows XP. Feedback was provided using a CRT monitor with 100 Hz update. This system acquired data from 32 channels at 1200 Hz using two g.tec g.USBamp amplifier/digitizer systems and a block size of 100 ms, i.e., signal display, signal processing, and feedback were updated 10 times per second. In the

Cursor Task signal processing/task configuration, BCI2000 extracted signal features from all 32 channels using an autoregressive method to calculate voltage spectra, and provided three-dimensional feedback using the Cursor Task application. In the *P3 Speller* signal processing/task configuration, BCI2000 extracted and classified average waveforms from all 32 channels, and implemented the row/column speller paradigm described in [4].

In the *Cursor Task* configuration, the mean processing latency of the BCI2000-based system was 20 ms. In other words, BCI2000 calculated the spectra of 32 channels, and translated the spectra into a three-dimensional movement signal, in 20 ms. The mean processing latency for the *P3 Speller* configuration was 8 ms. For both configurations, the average video output latency (i.e., the latency between the end of processing and update on the video screen) was 6.23 ms. Thus, while MS-Windows is not a real-time operating system and thus does not guarantee specific timing of events, in each test case, system latency easily satisfied the real-time requirements of BCI operation. Furthermore, processor load was sufficiently low to guarantee reliable operation. This indicates that BCI2000 could have handled even higher sampling rates, larger numbers of channels, or more complex signal processing methods. Please see [13] for a comprehensive description of the complete timing evaluation.

3.5 BCI2000 Implementations and Their Impact

We tested the adaptability and online performance of BCI2000 by using it to implement two very different BCI designs, each of which had previously been implemented only by its own highly specialized software/hardware system. In each case, the BCI2000 implementation required minimal effort to set up and yielded results comparable to those reported for the dedicated system. Furthermore, in each case the standard BCI data storage format readily supported the appropriate offline data analyses.

To implement mu/beta rhythm cursor control, we configured BCI2000 with autoregressive spectral estimation and the cursor task. In our configuration of this task, a target appeared in one of four possible locations along the right edge of the screen. Then, a cursor appeared at the left edge and moved from left to right at a constant rate with its vertical movement controlled by the power in a mu or beta rhythm frequency band at a location over sensorimotor cortex (see [8] for further detail). To date, several hundred people have used this system extensively (i.e., 4–300 sessions each). Figure 3.3-A illustrates the spectral and topographical features of the mu/beta rhythm control that users achieve and that allows them to move the cursor to the designated target.

To implement the P300-based BCI paradigm described by Donchin and his colleagues [3, 4], we configured BCI2000 with the temporal filter that averages evoked potentials and the P3 Speller application. As illustrated in Fig. 3.3-B, the results are similar to those reported for the original hardware/software P300 BCI system [4]. As described earlier in this book, many peer-reviewed publications have used BCI2000 using these or other configurations.

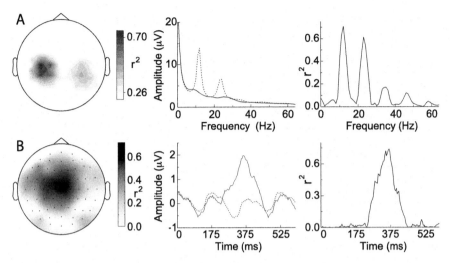

Fig. 3.3 BCI2000 implementations of common BCI designs (from [10]). **A**: Mu/beta rhythm control of cursor movement. *Left*: Topographical distribution on the scalp (nose on *top*) of control (measured as r^2 (the proportion of the single-trial variance that is due to target position)), calculated between top and bottom target positions for a 3-Hz band centered at 12 Hz. *Center*: Voltage spectra for a location over left sensorimotor cortex (i.e., C3 (see [12])) for top (*dashed line*) and bottom (*solid line*) targets. *Right*: Corresponding r^2 spectrum for top versus bottom targets. The user's control is sharply focused over sensorimotor cortex and in the mu- and beta-rhythm frequency bands. The data are comparable to those of earlier studies that used a specialized hardware/software system (e.g., [14]). **B**: P300 control of a spelling program. *Left*: Topographical distribution of the P300 potential at 340 ms after stimuli, measured as r^2 (calculated from averages of 15 stimuli) for stimuli including versus not including the desired character. *Center*: The time courses at the vertex of the voltages for stimuli including (*solid line*) or not including (*dashed line*) the desired character. *Right*: Corresponding r^2 time course. The data are comparable to those of earlier studies using a dedicated hardware/software system (see [3, 4]). Stimulus rate was 5.7 Hz (i.e., one every 175 ms). See text and references for other details

This chapter provided an introductory overview of the basic concepts of BCI2000, and how these concepts can be used to realize different BCI implementations. The following chapter provides an introductory hands-on tour through the BCI2000 platform.

References

1. Bayliss, J.D.: A flexible brain–computer interface. PhD thesis, University of Rochester, Rochester (2001). http://www.cs.rochester.edu/trs/robotics-trs.html
2. Bianchi, L., Babiloni, F., Cincotti, F., Salinari, S., Marciani, M.G.: Introducing BF++: a C++ framework for cognitive bio-feedback systems design. Methods Inf. Med. **42**(1), 102–110 (2003)
3. Donchin, E., Spencer, K.M., Wijesinghe, R.: The mental prosthesis: assessing the speed of a P300-based brain–computer interface. IEEE Trans. Rehabil. Eng. **8**(2), 174–179 (2000)

4. Farwell, L.A., Donchin, E.: Talking off the top of your head: toward a mental prosthesis utilizing event-related brain potentials. Electroencephalogr. Clin. Neurophysiol. **70**(6), 510–523 (1988)
5. Guger, C., Schlögl, A., Neuper, C., Walterspacher, D., Strein, T., Pfurtscheller, G.: Rapid prototyping of an EEG-based brain–computer interface (BCI). IEEE Trans. Neural Syst. Rehabil. Eng. **9**(1), 49–58 (2001)
6. Kemp, B., Värri, A., Rosa, A.C., Nielsen, K.D., Gade, J.: A simple format for exchange of digitized polygraphic recordings. Electroencephalogr. Clin. Neurophysiol. **82**(5), 391–393 (1992)
7. Mason, S.G., Birch, G.E.: A general framework for brain–computer interface design. IEEE Trans. Neural Syst. Rehabil. Eng. **11**(1), 70–85 (2003)
8. McFarland, D.J., Neat, G.W., Wolpaw, J.R.: An EEG-based method for graded cursor control. Psychobiol. **21**, 77–81 (1993)
9. Renard, Y., Gibert, G., Congedo, M., Lotte, F., Maby, E., Hennion, B., Bertrand, O., Lecuyer, A.: OpenViBE: an open-source software platform to easily design, test and use Brain–Computer Interfaces. In: Autumn School: From Neural Code to Brain/Machine Interfaces (2007)
10. Schalk, G., McFarland, D., Hinterberger, T., Birbaumer, N., Wolpaw, J.: BCI2000: a general-purpose brain–computer interface (BCI) system. IEEE Trans. Biomed. Eng. **51**, 1034–1043 (2004)
11. Schlögl, A.: GDF – a general dataformat for biosignals (2009). http://arxiv.org/abs/cs.DB/0608052
12. Sharbrough, F., Chatrian, G., Lesser, R., Luders, H., Nuwer, M., Picton, T.: American electroencephalographic society guidelines for standard electrode position nomenclature. Electroencephalogr. Clin. Neurophysiol. **8**, 200–202 (1991)
13. Wilson, J.A., Mellinger, J., Schalk, G., Williams, J.: A procedure for measuring latencies in brain–computer interfaces. IEEE Trans. Biomed. Eng. (in press)
14. Wolpaw, J.R., McFarland, D.J., Neat, G.W., Forneris, C.A.: An EEG-based brain–computer interface for cursor control. Electroencephalogr. Clin. Neurophysiol. **78**(3), 252–259 (1991)
15. Wolpaw, J.R., Birbaumer, N., McFarland, D.J., Pfurtscheller, G., Vaughan, T.M.: Brain–computer interfaces for communication and control. Electroencephalogr. Clin. Neurophysiol. **113**(6), 767–791 (2002)

Chapter 4
Tour of BCI2000

In this chapter, we present a general overview of BCI2000 to demonstrate important features of the software and other concepts that are relevant to any BCI session. We will cover methods for starting the BCI2000 software; give a brief tour of the user interface; describe setting, saving, and loading parameters; and retrieving data offline.

4.1 Starting BCI2000

There are two methods for starting the BCI2000 software system. BCI2000 consists of four programs (i.e., modules) that need to be started up in a certain order. These modules handle acquisition of brain signals (i.e., Source module), processing of these brain signals (i.e., Signal Processing module), user feedback (i.e., User Application module), and the interface to the investigator (i.e., Operator module), respectively. These four modules can be started using the batch files in the BCI2000/batch directory. Alternatively, BCI2000 comes with a program called BCI2000Launcher that allows you to manage module startup using a graphical user interface.

4.1.1 Batch Files

A group of pre-configured batch files are located in the BCI2000/batch directory. In general, these may need to be modified to correspond with the specific modules used for a particular experiment, although in most instances the appropriate configuration already exists if g.tec amplifiers are used. In any batch file, the four modules are started with the start command, starting with the Operator, i.e., start operat.exe. Next, the Source, Signal Processing, and Application modules are started, with the IP address of the Operator passed as an argument. In most cases, this is on the local machine, which has an address of 127.0.0.1.

G. Schalk, J. Mellinger, *A Practical Guide to Brain–Computer Interfacing with BCI2000*,
© Springer-Verlag London Limited 2010

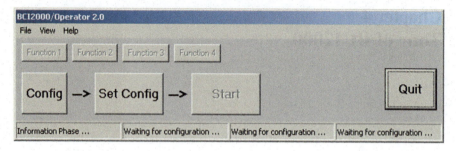

Fig. 4.1 The Operator main window

4.1.2 BCI2000 Launcher

The BCI2000Launcher program provides a convenient user interface for organizing BCI2000 applications and parameter files. The Source, Signal Processing, Application, and Operator modules are selected from a list, multiple parameter files can be passed to BCI2000 and loaded automatically, and all are launched simultaneously, replacing the need for multiple batch files.

To start the tour, perform the following steps:

Start BCI2000

1. Navigate to the BCI2000/batch directory.
2. Double-click the file CursorTask_SignalGenerator.bat. This will launch the modules required for a simulation of a BCI session based on sensorimotor rhythms.

The Operator window will appear, with all of the modules connected (Fig. 4.1).

4.2 Getting Help

When the main Operator window appears, choose *BCI2000 Help* from the *Help* menu. The Operator module's help page will be opened in a web browser window. On the left side of the help page, a navigation menu provides access to available help.

Further information about BCI2000 can be found on the BCI2000 Wiki, which is located at http://doc.bci2000.org/wiki/; and users can ask questions in the BCI2000 forums, which is located at http://bbs.bci2000.org. In these forums, the BCI2000 developers and other users can answer questions that are not covered in this book.

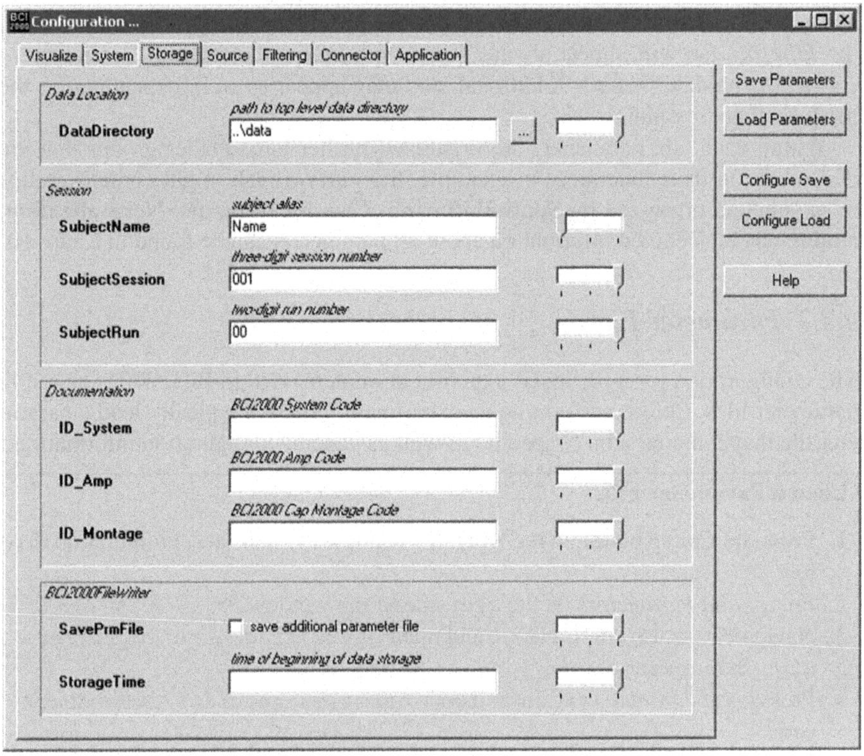

Fig. 4.2 The BCI2000 Configuration dialog

4.3 Configuring BCI2000

BCI2000 has an extensive set of configuration parameters, which are dependent on the particular modules being used, and must be set for each session. This section provides a review of configuring BCI2000, and loading and saving your settings for future use.

Enter the BCI2000 Configuration Dialog

1. Press the *Config* button in the Operator window.

4.3.1 Module Settings

When you press *Config*, the BCI2000 Configuration window appears (Fig. 4.2). Across the top of this window, you will see a series of several tabs. Each tab contains a different group of parameters; these parameters are specified by the (Source, Signal Processing, and User Application) modules that are being used. There will

always be *Visualize, System, Storage, Source, Connector,* and *Application* tabs;
the *Filtering* tab will almost always be present, unless the `DummySignalPro-`
`cessing` module is used. Additional tabs may appear as well, depending on the
modules that are running.

Within each tab, parameters are organized further into smaller groups that are
dependent on their function. For example, the *Filtering* tab might contain groups
of parameters organized by **SpatialFiltering**, **Classification**, and **Normalization**,
among others. Detailed information about all parameters can be found in Chap. 10.

4.3.2 Parameter Files

The configuration for a particular experiment implemented in BCI2000 is stored in
parameter files. Before performing an experiment, you will typically load a parameter file that contains subject-specific as well as general configuration information.

Load a Parameter File

1. Press the *Config* button in the Operator window to enter the Configuration dialog.
2. Press *Load Parameters* on the right side of the window.
3. Navigate to `BCI2000/parms/`, and open `parms/fragments/amplifiers/`
   `SignalGenerator.prm`.
4. Press *Load Parameters* again, and open `parms/mu_tutorial/MuFeedback.`
   `prm`.

Parameter files may contain all parameters needed for a particular experiment, or
only a subset. In the latter case, we call those parameter files *parameter fragments*.
Each fragment can contain information that can be reused across sessions and subjects, e.g., a fragment might contain only those settings that pertain to the cursor
task and that can be applied to any subject; another fragment might contain settings
for a particular amplifier; and yet another fragment might contain information for
an individual subject. Therefore, several fragments may be used together to define
the configuration of a particular experiment.

4.3.3 Parameter Help

It is possible to obtain information about each parameter within the Configuration
window. To do so, press the *Help* button on the right side of the Configuration window. This will turn the cursor into a question mark. If you click on a parameter label,
a web browser will appear with a help page that describes that parameter.

Parameter Help Example

1. Press the *Storage* tab.
2. Press the *Help* button on the right side of the Configuration window.
3. Use the mouse to click on the **SubjectName** parameter.
4. Information about the **SubjectName** parameter is shown.

4.4 Important Parameters

The BCI2000 framework will always include a number of parameters, regardless of the type of experiment. In this section, the parameters that define the location and file naming conventions of the data file will be configured.

Setting the Storage Parameters

1. Open the Configuration window, and navigate to the *Storage* tab.
2. Under **DataDirectory**, the value corresponds to the base directory in which all acquired data will be saved.
3. Relative or absolute paths can be used for this value. Relative paths are relative to the BCI2000/prog/ directory. In this case, ../data is specified, so data will be saved in BCI2000/data/.
4. The **SubjectName** parameter should be set to a value such as TEST, or the subject's initials. Set this to test for this demo.
5. The parameter **SubjectSession** should be set to a three-digit value corresponding to the number of sessions that a particular subject has participated in. Set this to 001 for this demo.
6. The parameter **SubjectRun** should be set to 00 at the start of a new session. This number will increment automatically with each run, i.e., existing data files will never be overwritten if the **SubjectRun** value equals that of an existing file.

4.5 Applying Parameters

Once you have defined the experimental parameters, they must be applied to BCI2000.

Applying Parameters

1. Close the Configuration window by pressing the "X" button located on the top-right corner of the window. This will accept your parameter changes.
2. In the main Operator window, click *Set Config*. This will apply the changes in configuration.
3. The Operator module will send the configuration parameters to the other BCI2000 modules, which in turn validate the values to ensure that they are consistent.
4. If all parameter values are consistent, the signal source window will appear and display the brain signals.
5. The signal source window can be moved and resized; this position is preserved across BCI2000 sessions.

In a true BCI session, in which you will be recording signals from the brain, you will use the signal source window to assess the quality of the recorded signals. In this simulation session, you will notice a change in the simulated brain signal when you move your mouse. We will use these signal changes to control a cursor on the screen.

When you right-click on the signal window, you will notice a context menu with particular display options, such as increasing/decreasing the number of channels displayed, choosing display colors, applying filters to the signal, etc. Please see Sect. 10.1.5 for more details.

4.6 Starting a Run

In this demonstration, you will simulate a real cursor movement experiment by modulating artificial brain signals with your mouse.

Performing a Session

1. Perform all of the steps described above.
2. Press *Set Config* to apply parameters.
3. You will see the simulated brain signals on the screen. Channels 1 and 2 contain 10 Hz sinusoidal signals whose amplitudes are modulated by the position of the mouse.
4. To increase or decrease the amplitude of channel 1, move the mouse up and down on the screen, respectively. To increase or decrease the amplitude of channel 2, move the mouse left and right on the screen, respectively.
5. Press the *Start* button to start the experiment, i.e., a simulated feedback session.
6. During the session, you will see a cursor that moves from left to right on the screen at a constant rate.
7. You will also see one of two targets on the screen. Your goal is to direct the cursor to the target.
8. The vertical velocity is controlled by the amplitude of channel 1. To move the cursor up, increase the amplitude, and to move it down, decrease the amplitude.
9. When the run has completed, the *Suspend* button will change back to *Start*.

The period during which the cursor moves from left to right is called a trial. During operation, there will be a number of successive trials. Many trials make up one experimental run, where each run usually lasts about 3–5 minutes. When the run is finished, BCI2000 will stop operation on its own, and the *Start* button will change to *Resume*. At that time, the corresponding data file has been closed. Clicking *Resume* each time, you may add as many runs to the current session as you like. All runs within a session will be stored as separate files in that session's data directory.

BCI2000 contains a component that adapts itself to the brain signal's characteristics (i.e., its mean value and its amount of variation). This component needs to observe the signal for a few trials until it has appropriately adapted to those characteristics. This means that initially you may not feel a correspondence between your actions and the cursor. As you proceed through a few trials, you will note that control becomes more accurate. Also, it is important that you place the mouse cursor in a center position at the beginning of each trial to give you enough space to move in the required direction. Likewise, when a real subject controls the cursor using brain signals, the system will need a few trials to adapt to the subject's general signal characteristics. (It may take the subject quite some time to learn how to actually control that signal.) When you have acquired one or more runs, you may quit BCI2000.

4.6.1 Data Storage

All data acquired during a run are stored in a single BCI2000 data file that has the extension .dat. This data file contains all brain signals, all parameters you just defined, and all event markers (i.e., state values) that encode important events during each run. The location and name of the data file depends on the *Storage* parameters. The file naming convention based on these parameters is: <DataDirectory>/<SubjectName>S<SubjectSession>/ <SubjectName>S<SubjectSession>R<SubjectRun>.dat, e.g., data/ testS001/testS001R00.dat.

4.7 Retrieving Data Offline

4.7.1 Reviewing Brain Signals and Event Markers

Several methods for retrieving data offline for viewing or analysis exist. The BCI2000Viewer program allows you to review brain activity and associated event markers.

Viewing Brain Signals

1. In Windows Explorer, navigate to the BCI2000Viewer directory i.e., BCI2000/tools/BCI2000Viewer, and double-click the BCI2000Viewer.exe file to start the program.
2. In BCI2000Viewer (see Fig. 4.3), open the data file that was created for this demo session, located at BCI2000/data/testS001/testS001R00.dat.

Every BCI2000 data file contains several experiment-specific event marker channels that are stored with every signal sample.

View Event Markers

1. The right side of the BCI2000Viewer program contains a list of channels and event markers, which are called states in BCI2000.
2. To view particular event markers, click the box next to the corresponding state name.
3. Check the **TargetCode**, **ResultCode**, and **Feedback** checkboxes.

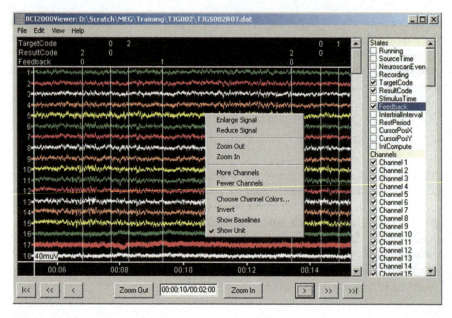

Fig. 4.3 The BCI2000Viewer program

The values of these event markers will be displayed on top of the signal traces. Whenever **TargetCode** changes from 0, a target was displayed during the recording. Whenever **ResultCode** changes from 0, a target has been hit by the cursor. A **Feedback** value of 1 indicates that the feedback cursor was visible. In data analysis, this state information makes it possible to determine trial structure and data labels. It is important to note here that the number, names, and meaning of the state variables are not predefined, but rather depend on the experiment, i.e., the particular collection of Source, Signal Processing, and User Application modules used for that experiment. Thus, you can easily write BCI2000 code that would add several new state variables with new meaning. These state variables would automatically be included in the data file and appear in BCI2000Viewer.

Help on state variables is available as well. Right-click the **TargetCode** entry in the top right corner of the BCI2000Viewer window and choose *Help* on the **TargetCode** state variable from the context menu. A browser window will open up and will show a help page that describes that variable.

4.7.2 Viewing Parameters with BCI2000FileInfo

In addition to state information, each data file contains the full set of parameters and their values that defined recording of this particular experimental run. You may view those parameters, or even save them to a parameter file in original or modified form, using the `BCI2000FileInfo` program that is located in `tools/BCI2000FileInfo`.

To open a `.dat` file in `BCI2000FileInfo`, drag and drop it onto the program symbol or into the program's main window. Then, click *Show Parameters*. This will open a parameter configuration dialog identical to the one you know from the Operator module.

4.7.3 Interactions with External Software

BCI2000 provides several software components that support interactions with external software, in particular with Matlab, FieldTrip (through a contribution by Robert Oostenveld) and Python (through a contribution by Jeremy Hill). These components can be used to load data from BCI2000 data files to analyze data offline, or to implement and test new signal processing techniques or feedback paradigms online. BCI2000 also provides a simple network-based interface (called the AppConnector protocol) that any external software can use to interact with BCI2000 in real time. A selection of these methods is covered in Chap. 6.

4.8 Overview of Additional Application Modules

BCI2000 contains several core application modules that are capable of providing multiple feedback methods, including visual and audio presentation. You have already been introduced to the cursor task; the remaining core applications are briefly covered in the following sections.

4.8.1 Stimulus Presentation

In addition to the cursor task, BCI2000 comes with a multi-purpose stimulus presentation program. This module has many of the features of dedicated programs like Presentation or e-Prime. The difference between these dedicated programs and stimulus presentation in BCI2000 is that BCI2000 makes it very easy to integrate that stimulus presentation with real-time data acquisition, processing, and feedback.

In BCI research, the BCI2000 Stimulus Presentation module is often used for an initial mu rhythm session. Because this module has good timing characteristics, it is also suitable for a wide range of psychophysiological experiments such as the ERP experiments described here. Furthermore, it can be used in conjunction with the P3 Signal Processing module (which can average and classify ERPs) to provide real-time feedback to evoked potentials.

To try the Stimulus Presentation module, start BCI2000 using the `StimulusPresentation_SignalGenerator.bat` file in the `batch` directory. Then, click *Config*, and load the configuration file at `parms/examples/StimulusPre-sentation_SignalGenerator.prm`. Navigate to the *Storage* tab and enter a subject ID into the **SubjectName** parameter. Run the experiment by clicking *Set Config* and *Start*. (Note: The goal here is simply to introduce you to the Stimulus Presentation module. We will not discuss all possibilities of this module here. Please see Sect. 10.8.2 for more information.) When you are done, locate the resulting data file at `data/<SubjectName>001/<SubjectName>S001R01.dat`, and open it using `BCI2000FileViewer`, similar to what you did for the simulated cursor movement session earlier in this tour. In `BCI2000FileViewer`'s main window, check the **StimulusCode** checkbox to display the **StimulusCode** state variable. During presentation of a stimulus, the **StimulusCode** state is set to an ordinal number that corresponds to which stimulus was presented. In data analysis, this information may then be used to segment data into epochs and to group epochs according to different stimuli.

4.8.2 P300 Speller

As part of its core distribution, BCI2000 also comes with the `P300Speller` module. The P300 Speller implements a BCI that uses evoked responses to select items from a rectangular matrix, a paradigm originally described by Farwell and Donchin.

To calibrate the P300 Speller for an individual subject, it is first operated in a "copy spelling" mode, which prompts the user to pay attention to pre-defined letters in sequence. Start `batch/P3Speller_SignalGenerator.bat`, click *Config*, and load `parms/examples/P3Speller_CopySpelling.prm`. Close the configuration window, then click *Set Config* and *Start* to view copy spelling in action. In this example, the simulated EEG will simulate evoked responses to the correct stimuli (i.e., correct row and column). This makes it possible to test signal classification and spelling functionality. In this example, spelled letters will correspond to the letters to be copied.

In real experiments, such as those described later in this book, the experimenter would use data from the copy spelling session to calibrate the P300 Speller. Once calibrated, the purpose of the P300 Speller is to enable the user to choose letters from the matrix, i.e., "free spelling," without a pre-defined letter sequence. The speller also supports multiple matrices (or "menus"), graphical icons, and auditory stimuli (i.e., wave files). To perform a "free spelling" demo with multiple menus, click *Config* and load `parms/examples/P3Speller_Menus.prm`. Then close the configuration window, click *Set Config*, and *Start*. In simulation mode, you can select a matrix element by clicking it with the mouse. Actual P300 classification is done by averaging several responses to a particular stimulus. Thus, clicking an item will not immediately select it. Rather, mouse selection will override classification once the number of averaging epochs is reached. This will take a few seconds.

4.9 Where to Go from Here

In this tour, we covered several basic aspects of BCI2000 theory and operation. This included a brief introduction to different BCI2000 modules that support BCI operation using sensorimotor rhythms and evoked responses. The following chapter provides tutorials for implementing working BCI systems using each of these two brain signals.

Chapter 5
User Tutorials

5.1 General System Configuration

Prior to running experiments, the computer and display system must be configured for both the experimenter and the subject. This tutorial assumes that you will be using a dual-monitor setup as shown below, with the experimenter of the sessions operating on monitor 1, and the subject watching on monitor 2. This system configuration is typical for most BCI experiments, and can be used for the Cursor movement, P300 Speller [1], or Stimulus Presentation tasks.

Configure a Dual-Monitor Setup

1. Open *Display Properties* by right clicking on an empty portion of the desktop and clicking *Properties*.
2. Navigate to the *Settings* tab.
3. You will see a representation of the position and geometry of both monitors. Ensure that monitor 2 is enabled by clicking on it, and clicking the check box labeled **Extend my Windows desktop to this monitor**, and hitting *Apply*.
4. First, make sure that monitor 2 is aligned with the top of monitor 1 by clicking and dragging it to the correct position (Fig. 5.1).
5. Make note of the width of monitor 1 and the resolution of monitor 2. In this example, monitor 1 is 2,048 pixels wide, and monitor 2 is 1,024 pixels wide by 768 pixels tall (Fig. 5.2).
6. Close the display windows.

5.2 Virtual Cursor Movement with the Sensorimotor Rhythms

A BCI cursor movement experiment is generally done in three steps: (1) obtaining sensorimotor rhythm parameters in a screening session; (2) analyzing the screening data and picking the best EEG features; (3) configuring BCI2000 with the chosen

© Springer-Verlag London Limited 2010

Fig. 5.1 Here we need to make sure that monitor *2* is aligned along the top with monitor *1*, and that monitor *2* is to the right of monitor *1*

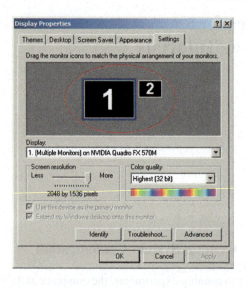

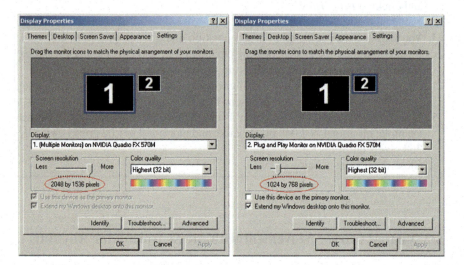

Fig. 5.2 What we need to make a note of is the **Width** of monitor *1*, and the **Width and Height** of monitor *2*. In this example we see that monitor *1* is 2,048 pixels wide, and monitor *2* is 1,024 pixels wide by 768 pixels tall

features; and (4) running an online session in which the subject moves a computer cursor. The following four sections will cover these four steps.

5.2.1 Obtaining Sensorimotor Rhythm Parameters in an Initial Session

Although the basic properties of mu/beta rhythms are identical for all humans, spatial patterns and exact frequencies are different from person to person. Thus, it is necessary to obtain these subject-specific parameters prior to any feedback experiments, i.e., to calibrate the BCI system using data acquired from an initial session.

In this initial session, the subject is instructed to imagine hand and/or foot movements in response to visual cues. To identify a subject's mu/beta rhythm, offline analyses then determine the frequency and location whose activity changes the most across conditions (e.g., hand imagery and rest). These analyses result in spectra calculated at different locations or in topographical plots at particular frequencies.

First, you need to connect the sensors to the amplifier system. For EEG, you may want to review Sect. 2.3 regarding electrode placement, etc. Then, you need to configure BCI2000 for this session as described below.

Configure the Mu/Beta Rhythm Session

1. Connect your amplifier to the computer, and turn it on.*
2. Run the `batch/StimulusPresentation_<amplifier>.bat` file. For example, if you are using the g.MOBIlab amplifier, you would run `batch/StimulusPresentation_gMOBIlab.bat`.
3. Press the *Config* button to bring up the BCI2000 Configuration window.
4. Press *Load Parameters* and load `parms/fragments/amplifiers/<amplifier>.prm`.
5. Next, load `parms/mu_tutorial/InitialMuSession.prm`.
6. In the *Storage* tab, set **SubjectName** to the subject's initials, set **SubjectSession** to `001`, and **SubjectRun** to `01`.
7. In the *Source* tab, set **ChannelNames** to the electrode names corresponding to each channel. In this example, `F3 F4 T7 C3 Cz C4 T8 Pz` are used, based on their respective positions in the 10–20 convention.
8. If using a g.MOBIlab amplifier, set **COM port** to the port name that you found earlier, e.g., `COM8:`.
9. In the *Application* tab, set **WindowWidth** to the width of the subject's display monitor, and **WindowHeight** to the height (`1024` and `768`, respectively, in this example).
10. Set the **WindowLeft** to the width of the experimenter's monitor. In this example, this would be `2048`.

Configure the Mu/Beta Rhythm Session (continued)

11. Take note of the **Sequence** field. This field contains four single-digit num-
    bers separated by one space. These numbers correspond to the presentation
    frequency of each row in the **Stimuli** matrix. Initially it will be 1 1 1 1, in-
    dicating that stimuli 1 through 4 should each be presented an equal number
    of times; the first number corresponds to the left hand, the second is the right
    hand, the third for both hands, the fourth for both feet. Therefore, setting this
    field to 2 1 0 1 will call for the subject to move the left hand twice as often as
    the right or both feet, and will never call for both hands to be moved.
12. Press *Save Parameters*, and save the file with a descriptive name. You can use
    this parameter file as the basis for future sessions.

** If your amplifier is from the gMOBIlab family, you will need to make a note of
the port it is connected to. In order to determine that port, go to the Windows Start
Menu, and choose* Start → Control Panel → System → Hardware → Device Manager → Ports (COM & LPT). *If you are using a different amplifier, please check
the amplifier documentation, and/or the amplifier-specific sections in Chaps. 10
and 11.*

During the initial session, the subject's screen will either be blank, or displaying
an arrow pointing up, down, left, or right. The instructions for the subject for each
of these periods are as follows:

Instructions to the Subject

1. When a left or right arrow is displayed, imagine movement of the respective
   hand. The imagined movement should be continuous opening and closing of
   the hand (e.g., squeezing a tennis ball) at a rate of about one opening and closing per second.
2. When an up arrow is displayed, imagine simultaneous movement of both
   hands. This should be the same kind of movement as described for a single
   hand.
3. When a down arrow is displayed, imagine movements of both feet. The movement should be similar to the one described for hands, i.e., imagine opening
   and closing your feet as if you could use them to grip an object.
4. When you see a blank screen, relax and stop any movement imagery.

5.2.2 Performing the Initial Sensorimotor Rhythm Session

To start an experimental run, click *Run* in the operator window. Each run gathers
20 data points, or "trials," that differentiate between moving the left hand, the right
hand, both hands, and both feet. Ideally, there should be 100 trials, meaning that
five runs are suggested. This is done as five separate runs instead of one to allow
the subject a chance between each run to rest, blink, swallow, speak, or have some
water if so desired.

5.2.3   Analyzing the Initial Sensorimotor Rhythm Session

To identify the parameters (i.e., frequency and location) of a subject's mu/beta rhythm, we will determine how different the EEG signal amplitude is at different frequencies and locations between rest and movement imagery, or between different types of imaginations. BCI2000 provides an "Offline Analysis" tool for this purpose.

5.2.3.1   Generating Feature Maps

The first step in these analyses is to separate data into amplitudes at individual frequencies and locations. For our purposes, these amplitudes are called features, and their correspondence with the subject's imagination will be plotted as a so-called feature map. From a feature map, it is possible to determine those frequencies and locations whose amplitude is maximally correlated with the subject's task, i.e., those features that are most different between two conditions. These features will subsequently be used to provide feedback in a BCI experiment.

To generate a feature map from the initial session's data, perform the following steps:

Generating a Feature Map

1. Start the BCI2000 "Offline Analysis" tool. If you have a version of Matlab installed, run `tools/OfflineAnalysis/OfflineAnalysis.bat`. Otherwise, download the MATLAB Component Runtime (MCR) from `http://www.bci2000.org/downloads/bin/MCRInstaller.exe` and install it. Then, run `tools/OfflineAnalysis/OfflineAnalysisWin.exe` to begin analyzing your data.
2. In the **Analysis Domain** field, choose **Frequency**.
3. In the **Acquisition Type** field, choose **EEG**.
4. As a **Spatial Filter**, choose **Common Average Reference (CAR)**.
5. Enter `states.StimulusBegin == 1` into the **Trial Change Condition** field.
6. Into the field labeled **Target Condition 1**, enter `(states.StimulusCode == 0)`.
7. Enter the word `Rest` into the field labeled **Target Condition Label 1**.
8. Similarly, enter `(states.StimulusCode == 2)` into the **Target Condition 2** field, and `Right Hand` into **Target Condition Label 2**.
9. Click the *Add* button located besides the **Data Files** field. A file chooser dialog will appear; navigate to `data/mu/<Subject>001`, and select all .dat files that you just collected (use your keyboard's *ctrl* button to click-select multiple files), then click the dialog's *Open* button.
10. Click *Generate Plots*, and wait for the feature map to appear.

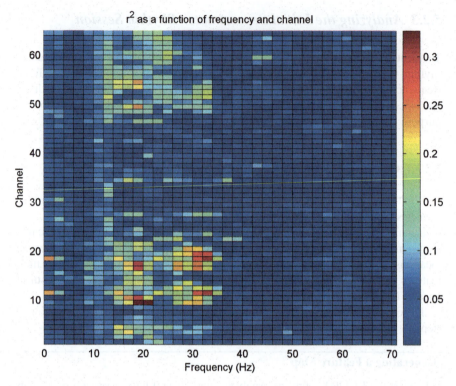

r² as a function of frequency and channel

Fig. 5.3 An example EEG feature map

Once the computation is complete, you will see a feature map similar to Fig. 5.3. In that plot, the horizontal axis corresponds to frequencies, and the vertical axis corresponds to individual channels. Color codes represent r^2 values, which are numbers between 0 and 1. r^2 values provide a measure for the amount to which a particular EEG feature (i.e., amplitude at a particular frequency and location) is influenced by the subject's task (e.g., hand vs. foot imagery).

Typically, there will be clusters of large r^2 values in the feature map. The initial step to configure the online system is to determine which brain signal feature differed the most between two particular tasks. At the same time, it is important to verify whether the properties of that feature are consistent with the mu/beta rhythm's known properties. This verification is necessary to avoid misconfiguration of the system to use EEG artifacts, other noise, or random effects, rather than features originating within the brain.

To verify candidate features, pick the four largest r^2 values from the feature map between 8 and 36 Hz, and read off their frequencies and channels. The program's "Data Cursor" tool (*Data Cursor* from the *Tools* menu) may be helpful for this. Then, enter the channel numbers you read off the feature map in the analysis program's **Spectra Channels** field, and enter the frequencies in the field for the **Topo Frequencies**. Then, click the *Generate Plots* button.

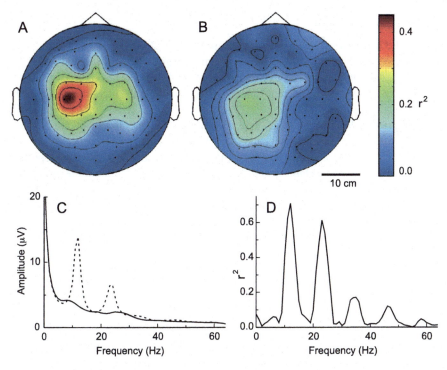

Fig. 5.4 Scalp topographies of activation during movements (modified from [2]). **A, B**: Topographical distribution on the scalp of the difference (measured as r^2, or the proportion of the single-trial variance that is due to the task), calculated for actual (**A**) and imagined (**B**) right-hand movements and rest for a 3-Hz band centered at 12 Hz. **C**: Example voltage spectra for a different subject and a location over left sensorimotor cortex (i.e., C3 (see [3])) for comparing rest (*dashed line*) and imagery (*solid line*). **D**: Corresponding r^2 spectrum for rest vs. imagery. Signal modulations are focused over sensorimotor cortex and in the mu- and beta-rhythm frequency bands

The generated *topography plots* display the spatial distribution of r^2 values similar to those in Fig. 5.4-A/B. In particular, there should be a clear maximum of r^2 values over the left motor cortex as shown in subfigure A for actual movements and B for imagined movements. The generated *spectra plots* display amplitude distributions and r^2 values across a range of frequencies. Ideally, they should appear similar to those in Fig. 5.4-C/D.

5.2.3.2 Analyzing Remaining Conditions

Up to now, you performed an analysis that determined how brain activity is related to imagined movements of the right hand. In order to choose the most useful channel and frequency for online feedback, perform similar analyses for the remaining conditions:

Remaining Conditions

1. In the analysis program's **Target Condition 2** field, enter `states.Stim-`
   `ulusCode == 1`, and `Left Hand` into **Target Condition Label 2**.
2. Make sure the **Overwrite existing plots** check box is unchecked.
3. Click *Generate Plots* to create a feature map for imagined movement of the left
   hand.
4. As previously, pick the four largest r^2 values, and compute spectra and topographies for their channels and frequencies.
5. Results should somewhat resemble that derived for the right hand, except that
   the colored activity changes should appear over the right and not the left motor
   cortex.
6. Repeat the analysis for conditions `states.StimulusCode == 3` (Both
   Hands), and `states.StimulusCode == 4` (Both Feet).
7. For the `both hands` condition, the result should resemble a combination of
   `left hand` and `right hand` results.
8. For `both feet`, modulated activity should be centered around electrode Cz.

5.2.3.3 Picking Optimal Features

You should now have identified a number of candidates of channel/frequency pairs
for each condition, and you should have some impression how physiologically plau-
sible these are. In particular, we would like to see results that at least somewhat resemble those shown in Fig. 5.4, i.e., frequencies with the highest r^2 value between
8 and 36 Hz, and locations with the highest r^2 value at appropriate locations (e.g.,
C3 for right hand movement/imagery; C4 for left hand movement/imagery; C3 and
C4 for both hands; and Cz for both feet).

5.2.4 Configuring Online Feedback

After picking a desired target feature (i.e., desired frequency and location), we can
now configure BCI2000 to extract and use that feature.

Configuring BCI2000

1. Run the `batch/CursorTask_<amplifier>.bat` file. In the example, we
   are using the gMOBIlab amplifier, so we would run `batch/CursorTask_`
   `gMOBIlab.bat`.
2. Press the *Config* button to bring up the BCI2000 Configuration window.
3. Press *Load Parameters*, and first load the parameter file created and saved for
   the screening session. This will re-load many parameters that can be re-used.
4. Next, press *Load Parameters* and load: `parms/fragments/amplifiers/`
   `<amplifier>.prm`, then `parms/mu_tutorial/MuFeedback.prm`.

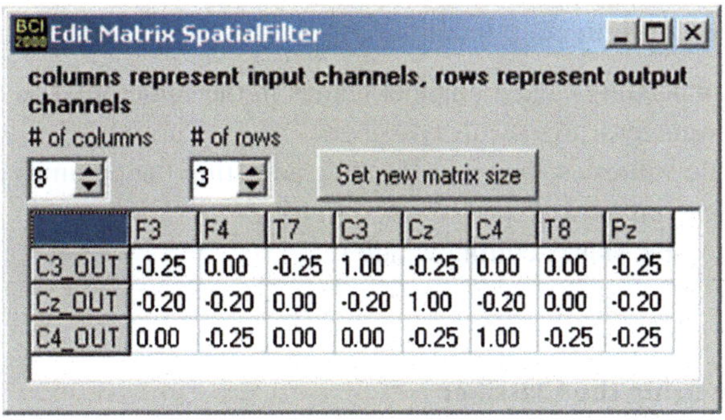

Fig. 5.5 Spatial Filter configuration

Configuring BCI2000 (continued)

5. In *Storage*, change **SubjectName** to the subject's initials.
6. Set **SubjectSession** to 002 and **SubjectRun** to 01.
7. Press *Save Parameters*, and save to an appropriate file.

This file is now the base parameter file for your amplifier when configuring your system to user-specific settings for the cursor task portion of future sessions.

5.2.4.1 Configuring the Spatial Filter

The Spatial Filter computes a weighted combination of the incoming data from the electrodes based on their placement on the scalp of the subject. Because we are targeting specific areas of the brain to monitor, we use a spatial filter that allows the program to identify when the electrode of interest is activating specifically. This is done by subtracting the average of the surrounding electrodes' data from the electrode of interest. For example, as seen in Fig. 5.5, in which the incoming signals are in the columns and in which the rows represent the output of the spatial filter, the output channel C3_OUT is the data from electrode C3 minus one-quarter of each of electrodes F3, T7, Cz, and Pz. Such a filter is called a "Laplacian filter." Please note that you should change the column labels depending on the montage and/or channel order you used.

Spatial Filter Configuration

1. Enter the Configuration window.
2. Press the *Filtering* tab.
3. Set the **SpatialFilterType** to **Full**.
4. Press the *Edit Matrix* button next to the **SpatialFilter** parameter.
5. Set **# of Columns** to 8, and **# of Rows** to 3.
6. Right-click the empty top-left matrix entry, and choose "Edit labels" from the context menu.

Spatial Filter Configuration (continued)

7. For column headings, enter channel names in the same order as the **Channel-Names** parameter.
8. Set the table values as shown in Fig. 5.5, adjusting the columns depending on the montage you used.
9. Close out of the Matrix editing window.

5.2.4.2 Configuring the Classifier

After defining an appropriate filter, you have to tell BCI2000 which location and frequency should be used for the feedback session. This is done as follows:

Linear Classifier Configuration

1. In the Configuration window, select *Edit Matrix* next to the **Classifier** parameter in the *Filtering* tab.
2. Set **Number of columns** to 4, and **Number of rows** to 1 (or the number of features that you wish to use). Click *Set new matrix size* to apply your changes.
3. In the first column (of the first row), labeled **input channel**, enter the code for the channel you would like to use, e.g., C3_OUT, C4_OUT, or Cz_OUT.
4. If both hands are used, set **Number of rows** to 2, and click *Set new matrix size*. Enter C3_OUT in row 1, column 1, and enter C4_OUT in row 2, column 1.
5. In the second column, labeled **input element (bin)**, enter the frequency of your feature, immediately followed with Hz, e.g., 12Hz.
6. In the third column, enter the value 2. This value defines the output control signal that the resulting signal should go in. In the cursor movement task, control signals 1–3 correspond to horizontal, vertical, and depth movement, respectively.
7. In the fourth column, enter −1 (minus one) as the weight.
8. Finally, save your configuration in a parameter file wherever you find appropriate.

It is important to note that this tutorial gives a functioning, but only simple, example. BCI2000 could be configured to use more complex spatial filters, perform different frequency estimation, or more complex classification. Please see Sect. 10.6 for more details on the configuration of the signal processing component.

5.2.5 Performing a Sensorimotor Rhythm Feedback Session

5.2.5.1 Starting the Session

To perform a mu/beta rhythm feedback session, start the BCI2000 system using the appropriate batch file at batch/CursorTask_<amplifier>.bat, or the link to that file which you created on the desktop. Then, load the configuration file that you saved for this subject previously by pressing *Config*, and then *Load Parameters*. Click *Set Config* to view the EEG signal, and prepare the subject for EEG recording as you did in the initial session.

5.2.5.2 Instructions to the Subject

When the subject is ready for EEG acquisition, it is time to brief the subject about the experimental task. Suggested instructions for the experimenter and subject are listed below.

Instructions to the Subject

1. After pressing the *Set Config* button, a screen with a yellow grid is initially presented.
2. Instruct the subject to minimize: (1) contraction of the muscles of the face/head, and swallowing; (2) eye blinks and eye movements; and (3) other motion.
3. As soon as the subject is ready and the EEG traces have stabilized, the investigator will start acquisition.
4. After pressing *Start*, there will be a brief pause of about 2 s (or equal to the **PreRunDuration** parameter).
5. A target will appear on the right edge of the screen for about one second. This is called the pre-feedback phase.
6. A cursor will appear on the left edge of the screen, and begins to move horizontally towards the right edge of the screen. Its vertical position is controlled by the EEG features that were defined in the previous step.
7. The subject's task is to move the cursor vertically so that it hits the target when the cursor hits the right edge of the screen. The time needed for the cursor to move from the left to the right edge of the screen should be approximately three seconds.
8. If the subject successfully hits the target, the target changes its color. Otherwise, no change occurs. In either case, this period lasts one second.
9. The screen will then turn black for one second. This indicates the end of the trial. After this one-second period, the next trial starts.

When a target is presented, the subject should imagine the type of movement associated with the channel-frequency features chosen for feedback. For example, when the largest r^2 value was associated with left hand movement imagination in the offline analysis, ask the subject to use this imagination to control the direction of cursor movement. Imagined movement will move the cursor upward on the screen, and relaxing will move it downward.

Explicitly specifying a type of imagination to control cursor movement will help the subject achieve initial cursor control. Once the subject has become more proficient with the task, motor imagery typically becomes less important. In this situation, it is not uncommon that subjects report that they just "imagine moving the cursor."

Provided that subjects are asked to minimize artifacts, he/she should be further assisted in these efforts by providing a comfortable chair and a dimly lit room. The experimenter must carefully monitor the EEG and alert the subject in the case he/she has forgotten some of the instructions. When the experimenter is sure that his/her instructions have been well understood, the recording session may start.

Click the *Start* button to start the feedback experiment. During the experiment, the subject's performance is written into a log window on the experimenter's screen, and recorded into a log file that is saved to disk in the session directory. The experimenter should minimize noise in the room and not disturb the subject.

5.2.5.3 Monitoring the Recording

After recording has started, the experimenter may feel the temptation to leave the subject alone during the run since most of the experimental activities are automated in BCI2000. However, the experimenter has several important tasks during the experiment.

Experimental Tasks for the Researcher
1. Fill in a run sheet to report information that is not automatically recorded by BCI2000 and that will later help when data are analyzed (e.g., subject did not seem to understand the instructions in the first run, instructions to the subject for a particular run, disturbances, etc.).
2. Monitor the EEG signal to verify the quality of the recording (e.g., no electrode contact failure, muscular, ocular, or motion artifacts, etc.).
3. Notify the subject if he/she is producing artifacts, keep the subject alert if getting drowsy, give the subject feedback about his/her performance so that interest, alertness, and attention is kept high.

The run sheet is extremely important, in that it documents important information or observations relevant to data analysis. In general, all information that could be important for data analysis and that is not explicitly documented elsewhere, need to be included. For example, the sampling rate is documented in BCI2000 parameters, while the name of the experimenter may not be documented elsewhere. Examples of such important details are comments about the subject's performance during a particular experimental run (e.g., subject getting sleepy, interruption at 12:34p), or about the experimental setup (e.g., used electrode cap #3, montage #2, distance of subject to screen 70 cm, etc.).

5.2.5.4 Multiple Sessions

Once a run has ended, BCI2000 goes into suspended state. Further runs will be added to the session when you click *Resume*. After the session has finished, you may want to save auto-adjusted parameters for the next session. Use *Save Parameters* from the configuration window to do this.

Alternatively, the *Load Parameters* dialog allows you to choose a data file rather than a parameter file, and thus use the configuration contained in a previous session's data file for the next session. However, parameters contained in a data file reflect the state at the beginning of the recording, so changes during a session's last run cannot be recovered that way.

When beginning the next session, don't forget to increment the **SessionNumber** parameter on the *Storage* tab. Otherwise, new runs will be added to the previous session's directory. As a safety net, BCI2000 will never overwrite existing data files, and it documents date and time in the **StorageTime** parameter. This allows the experimenter to later separate data files into runs even if the **SessionNumber** parameter has not been increased.

After each session, it is recommended that you analyze the recorded data in the same way as you did for the initial session. This allows you to track and adapt to signal changes in the subject's parameters that may occur in the course of learning.

5.2.5.5 Important Remarks

One critical element of such experiments is that they need to be consistent and rigorous. For example, a typical session will consist of a number (e.g., 4–8) of 3-minute experimental runs. Unless there is an obvious technical problem (e.g., the cursor always immediately jumps to the bottom of the screen, which would point to a misconfiguration of BCI2000), do not change any of the parameters (such as locations, frequencies, etc.) across these runs. When doing offline analyses, always strive to collect at least four runs with the exact same configuration. Because there is so much variability in the subject's performance and in the EEG, it is likely that you will otherwise not be able to derive meaningful results or conclusions. You may find that, for example, for three consecutive sessions the subject's best frequency is 12 Hz and not 10 Hz as initially configured. In this case, you could make this small adaptation to the parameters, and have a reasonable chance that it will actually improve the subject's performance.

5.3 P300 BCI Tutorial

The previous sections described the configuration and use of BCI2000 for sensorimotor (i.e., mu/beta) rhythms. The present chapter will describe a similar process for the P300 evoked potential. Just as with the mu rhythm, successful P300 BCI experiments require several steps that include obtaining initial data in a calibration session, selecting the best features related to the task, and finally performing the P300 spelling session. This section will cover all of these steps.

5.3.1 General System Configuration

The system should be configured for two monitors, as described in Sect. 5.1.

5.3.2 Obtaining P300 Parameters in the Calibration Session

Although the basic properties of the P300 evoked potential are the same for all individuals, the response's latency, width, and spatial pattern varies, and adaptation to individual parameters improves accuracy. Thus, it is necessary to obtain these individual parameters prior to performing spelling experiments.

Configure the P300 Spelling Session

1. Connect your amplifier to the computer, and turn it on.*
2. Run the `batch/P3Speller_<amplifier>.bat` file. For example, if you are using the g.MOBIlab amplifier, you would run `batch/P3Speller_gMOBIlab.bat`.
3. Press the *Config* button to bring up the BCI2000 Configuration window.
4. Press *Load Parameters* and load `parms/fragments/amplifiers/<amplifier>.prm`.
5. Next, load `parms/p3_tutorial/InitialP3Session.prm`.
6. In the *Storage* tab, set **SubjectName** to the subject's initials, set **SubjectSession** to `001`, and **SubjectRun** to `01`.
7. In the *Source* tab, set *ChannelNames* to the electrode names corresponding to each channel. In this example, `F3 F4 T7 C3 Cz C4 T8 Pz` are used, based on their respective positions in the 10–20 convention.
8. If using a g.MOBIlab amplifier, set **COM port** to the port name that you found earlier.
9. If a gMOBIlab is used, set the **COMport** parameter found earlier.
10. In the *Application* tab, set **WindowWidth** to the width of the subject's display monitor, and **WindowHeight** to the height (`1024` and `768`, respectively, in this example).
11. Set the **WindowLeft** to the width of the experimenter's monitor. In this example, this would be `2048`.
12. Press *Save Parameters*, and save the file as you deem fit.

* If your amplifier is from the gMOBIlab family, you will need to make a note of the port it is connected to. In order to determine that port, go to the Windows Start Menu, and choose Start → Settings → Control Panel → System → Hardware → Device Manager → Ports (COM & LPT).

This file is now the base parameter file for your amplifier when configuring the system to user-specific settings.

Fig. 5.6 P300 Speller matrix

5.3.2.1 Experimental Design

During the calibration session, the subject is asked to successively focus on different characters in a matrix (Fig. 5.6). During each run, the subject is asked to focus on the next letter in the word he or she is spelling, as the rows and columns flash randomly and successively so that sometimes the flashing corresponds to the column or row containing the target character and, more often, it will not. As the subject counts the number of times the next letter in the word flashes, a P300 response is generated. The purpose of the calibration session is to identify those features that discriminate between the desired and undesired rows and columns.

After the first few runs are collected, we will again use the "Offline Analysis" tool to determine which features (in this case, signals at a particular location and time after the stimulus) correspond to the row or column of the desired character.

5.3.2.2 Performing the Calibration Session

While the P300 response shares basic properties for all individuals, some specific details vary from person to person, such as latency, width, and spatial pattern. By adapting the basic parameters to the specifics associated with each individual subject, we can improve his or her accuracy drastically.

P300 Calibration

1. Start BCI2000 by running `batch/P3Speller_<amplifier>.bat`.
2. Press *Config*, and load the baseline parameters for copy spelling that you made earlier.
3. In the *Storage* tab, set **SubjectName** to the subject's initials, **SubjectSession** to `001`, and **SubjectRun** to `01`.
4. Press the *Application* tab.
5. Make sure that **InterpretMode** is set to `copy` mode, and **DisplayResults** (directly below InterpretMode) is unchecked.
6. Find the **TextToSpell** field. This should be set to `THE`, and you will be changing it after each run.
7. Press *Set Config* to apply this configuration.

P300 Calibration (continued)

8. Request that the subject sit in a relaxed position, and that the subject not move or speak during the runs.
9. Turning off or dimming the lights can improve subject focus and performance.
10. Showing the brain signal window on the subject's screen while describing EEG artifacts can assist in describing how artifact-generating behavior can be detrimental to the data.
11. Press *Start* to show the flashing character matrix, and describe what the subject is expected to do.
12. After you have explained the procedure, click *Suspend* to stop the process.
13. Delete that run of data, found at `data/P300/<Subject Initials>001/<Subject Initials>S001R01.dat`.
14. Press *Start* to record the run.
15. Once it has finished automatically, click *Config* and change the **TextToSpell** field in the **Application** tab to `QUICK`.
16. Press *SetConfig*.
17. Press *Start* to record the run.
18. Once it has finished automatically, click *Config* and change the **TextToSpell** field in the **Application** tab to `BROWN`.
19. Press *SetConfig* and then *Start* to record the run.
20. Once it has finished automatically, click *Config* and change the **TextToSpell** field in the **Application** tab to `FOX`.
21. Press *SetConfig* and *Start* to record the run.
22. Once this recording has finished, close BCI2000 and locate the saved data files for analysis.

5.3.3 Analyzing the Calibration Session with "Offline Analysis"

5.3.3.1 "Offline Analysis" Tool

We will now use the BCI2000 "Offline Analysis" tool to analyze the subject's initial session.

Offline Analysis

1. Run `tools/OfflineAnalysis/OfflineAnalysis.bat`.
2. In the **Analysis Domain** field, choose `Time(P300)`.
3. In the **Acquisition Type** field, choose `EEG`.
4. As a **Spatial Filter**, choose `Common Average Reference (CAR)`.
5. Set the **Trial Change Condition** field to `states.StimulusBegin == 1`.
6. Set **Target Condition 1** to: `(states.StimulusCode >0) & (states.StimulusType == 1)`.

Offline Analysis (continued)

7. Enter `Attended Stimuli` for **Target Condition Label 1**. "Attended Stimuli" refers to the letter or character the person is counting the flashes of, and triggers when the correct stimulus is shown.
8. Set **Target Condition 2** to: `(states.StimulusCode >0) & (states.StimulusType == 0)`.
9. Enter `Unattended Stimuli` for **Target Condition Label 2**. "Unattended stimulus" refers to the letters or characters the person is not counting the flashes of, and triggers when an incorrect stimulus is shown.
10. Click the *Add* button by the **Data Files** field.
11. In this new dialog, select all of the data files taken during this configuration session, and click *Open*.
12. Click *Generate Plots* and wait for the feature map to appear.

When this is complete, you will see a feature map similar to that in Fig. 5.7. In this plot, the vertical axis corresponds to the different locations, while the horizontal axis corresponds to the time after the stimulus. As before, the color shows the value of r^2 calculated between desired and not desired brain responses (i.e., responses to correct vs. incorrect row/columns). Red colors indicate a high correlation of the brain signal amplitude at the time/location with flashing of the desired row/column, and blue colors indicate little correlation. We are interested in finding large r^2 values between 200 and 550 ms.

Selecting Features

1. Pick the four points with the largest r^2 values between these times and record their time points and channels. The plot's *Data Cursor* tool (*Tools* Menu → *Data Cursor*) allows for discrete selection of time points.
2. In the example shown in Fig. 5.7, the four best data points have r^2 values `0.02218`, `0.02179`, `0.00328`, and `0.003`; occur at times `388.7ms`, `392.6ms`, `384.8ms`, and `365.2ms` respectively; and all four are detected by channel six.
3. To produce topographies and time course plots, you would enter 6 into the **Waveform Channels** field and `388.7, 392.6, 384.8, 365.2` into the **Topo Times** field.
4. Click *Generate Plots* to create the feature map again with a set of four graphs that show the correlation of the brain responses with the task category (i.e., desired vs. undesired row/column).
5. As seen in Fig. 5.8, the attended-stimulus reaction will typically be stronger than for the unattended stimuli, but in some cases the reverse is true. If the 'unattended' curve is larger than the 'attended' curve then make a note of this before moving on. The waveform seen here is similar to the others generated, only one is shown here for simplicity.

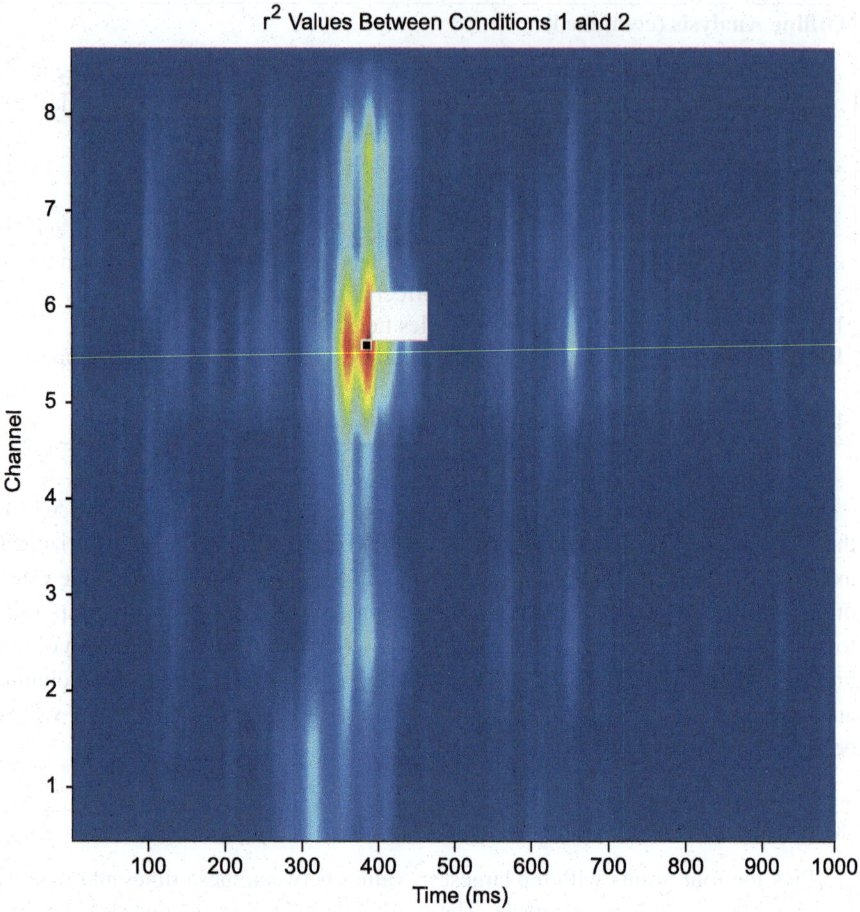

Fig. 5.7 Feature map produced by the "Offline Analysis" tool

Selecting Features (continued)

6. Additionally before moving on, determine the location of the response seen. The P300 response is generally observed centered on the Cz electrode, or just behind and directly in between the ears, and does not involve the frontal regions of the brain. Assuming these characteristics are present, it is proper to proceed.
7. If there are less than four points that seem appropriate, either because they are at the wrong electrodes, at the wrong times, or simply have too low of an r^2 value, that is fine; three or two values can work, though fewer values used will typically result in a lower accuracy.
8. Additionally, it may be helpful to re-run the analysis by setting the **Spatial Filter** to None, particularly when few channels are used.

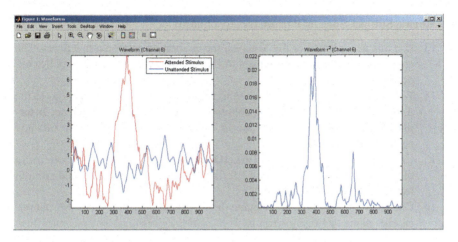

Fig. 5.8 Waveform plot produced by the "Offline Analysis" tool

Similar to what you did in the Mu Rhythm Tutorial, please note the location and time of the best features. Keep in mind that the P300 response generally peaks around the Cz/Pz electrodes, rather than in frontal or lateral regions of the brain. The next step is to configure BCI2000 to use the selected locations/times.

Storing Subject-Specific Parameters

1. Start BCI2000 using `batch/P3Speller_<amplifier>.bat` file.
2. Click *Config*, and load the configuration file saved previously.
3. Under the *Storage* tab, set the **SubjectSession** parameter to `002`, and the **SubjectName** parameter to the subject's initials.
4. Under the *Filtering* tab, click the *Edit Matrix* button next to **Classifier** near the bottom.
5. Change this matrix to have four columns and as many rows as values you are using obtained from the offline analysis, and click *Set New Matrix Size*.
6. In the first column, labeled **Input Channel**, enter the channel of the first value you use.
7. In the second column, labeled **Input Element (bin)**, enter the time of the best classification, immediately followed with `ms`, as in `388.7ms`.
8. In the third column, enter `1` as the output channel.
9. In the fourth column, enter `1` if the red line was larger than the blue line in the waveform plot (i.e., Fig. 5.8), and `-1` if the blue line was larger than the red line.
10. Repeat these steps for the remaining values, and close the matrix.
11. If a `Common Average Reference` spatial filter was used in the analysis, under the *Filtering* tab change the **SpatialFilterType** parameter to `Common Average Reference`; otherwise, leave it as `None`.
12. Click *Save Parameters* to save this file, naming it however you deem fit.

Storing Subject-Specific Parameters (continued)

13. Use this new parameter file to repeat the calibration session a few times. The goal of these repeated sessions is to increase the accuracy of the speller experiment.
14. When accuracy is at an acceptably high level (>90%), click *Config*, go to the *Application* tab, and delete the contents of the **Text to Spell** field, set **InterpretMode** to `online free mode`, and make sure the **DisplayResults** box is checked.
15. Click on *Edit Matrix* next to **TargetDefinitions** and scroll to the bottom. In the first column, replace 9 with `BS` (for "backspace"). In the second column replace 9 with `<BS>`.
16. Click *Save Parameters*, and change the `copy_spell` portion of this parameter file name to `free_spell`.
17. This parameter file is now ready to use for that specific subject for future P300 spelling experiments.

5.3.3.2 P300 Classifier

The use of the *Offline Analysis* program will familiarize the reader with the characteristics of the P300 response. At the same time, manually selecting the best features will typically result in reduced performance compared to when the features are automatically selected. Such automatic selection is performed using the *P300 Classifier* tool that is provided with BCI2000. This stand-alone program determines optimal features (i.e., signal times and channels) and corresponding weights automatically, and outputs those in a classifier matrix. Use of this program streamlines the configuration process at the expense of decreased hands-on experience with BCI data and the BCI2000 program itself.

For detailed instructions on using the P300 Classifier, see Chap. 10.11. To generate a parameter file containing optimized classification settings:

Using the P300 Classifier

1. Navigate to `BCI2000/tools/P300Classifier`, and double-click `P300-Classifier.exe`.
2. The P300 Classifier program appears.
3. In the top of the window, click on *Load Training Data Files*.
4. Select the data files collected from the four calibration runs, located at `BCI2000/data/P300/<Subject Initials>001/`.
5. To show the default settings, press the *Parameters* button. However, none of these settings should be changed right now.
6. Go back the *Data* pane, and click on *Generate Feature Weights*. To observe the progress, press the *Details* button.
7. On the *Details* pane, take note of how many sequences are required for 100% accuracy. This will be used later to configure the P300 spelling session.

Using the P300 Classifier (continued)

8. If less than 100% accuracy is achieved, your subject may not produce a strong P300 response, or you may need to collect more calibration data prior to proceeding.
9. When the classification is complete, click on *Write *.prm File* on the *Data* pane. Create a name for the parameter file, which will be used during the P300 spelling session later.

5.3.4 Performing a P300 Spelling Session

Now that the appropriate classification parameters have been determined, it is possible to perform a real spelling session. First, you should provide the subject with instructions that explain what they should expect, for example:

Subject Instructions

1. You will see a matrix of characters, numbers, and punctuation marks.
2. To choose a certain letter, concentrate on it by counting each time it flashes.
3. After some time, the computer determines from your brain signals which character you want and appends it to the text field located at the top of the window.
4. If the letter that appeared is not what you intended, concentrate on the "backspace" or "undo" field to remove it.
5. As in the previous sessions, please refrain from blinking, looking around other than at the letter you want to select, moving your head or body, speaking, or swallowing. These actions produce artifacts in the data that can throw off your selection.

The subject should be further assisted by providing a comfortable chair and a dimly lit room. The experimenter must carefully monitor the EEG and alert the subject in case he/she has forgotten some of the instructions. When the experimenter is sure that the instructions have been well understood, the recording session may start.

Starting a P300 Session

1. Open BCI2000 with `batch/P3Speller_<amplifier>.bat`.
2. Click on *Config*, then *Load Parameters*.
3. Load the base parameter file for your amplifier, and then the parameter file saved previously using either the `OfflineAnalysis` tool or the `P300 Classifier` tool.
4. If the P300 Classifier was used, in the *Filtering* tab, set **EpochsToAverage** to the ideal number found by how many flashes were needed to attain maximum accuracy. If the Offline Analysis tool was used, set this to 15.

Starting a P300 Session (continued)

5. In the *Application* tab, set **NumberOfSequences** to this number as well.
6. Delete the contents of the **Text to Spell** field.
7. Set **InterpretMode** to online free mode.
8. Make sure the **DisplayResults** box is checked.
9. Click on *Edit Matrix* next to **TargetDefinitions**, and scroll to the bottom.
10. In the first column, replace 9 with BS.
11. In the second column, replace 9 with <BS>.
12. Click *Save Parameters* and name as you see fit. Typically this parameter file would include the name of your amplifier with P300 and the subject's initials, along with Free_Spell. This file is now the file used for that specific subject in later P300 spelling sessions.
13. Click *Set Config* to view the EEG signal, and prepare the subject for EEG recording.
14. *Note*: When your latency value does not refer to an exact sample position, a warning message will be displayed once you click *SetConfig*. You may safely ignore that warning message for now.
15. On the subject's screen, a speller matrix the subject is already familiar with from the initial session is presented. However, in this session, no text is suggested; rather, the subject may choose freely which letters, words, and sentences to write.
16. Click the *Start* button to start the spelling experiment.

5.3.4.1 Multiple Sessions

Once a run has ended, BCI2000 goes into suspended state. Further runs will be added to the session when you click *Resume*. When starting the next session (typically, on a different day), don't forget to increment the **SessionNumber** parameter on the *Storage* tab. Otherwise, new runs will be added to the previous session's directory. BCI2000 will never overwrite existing data files but increment the largest run number that exists in a session directory. Moreover, it documents date and time in the **StorageTime** parameter. This allows the experimenter to later associate data files with multiple sessions by their time and date, even if the **SessionNumber** parameter has not been increased.

References

1. Farwell, L.A., Donchin, E.: Talking off the top of your head: toward a mental prosthesis utilizing event-related brain potentials. Electroencephalogr. Clin. Neurophysiol. **70**(6), 510–523 (1988)
2. Schalk, G., McFarland, D., Hinterberger, T., Birbaumer, N., Wolpaw, J.: BCI2000: a general-purpose brain–computer interface (BCI) system. IEEE Trans. Biomed. Eng. **51**, 1034–1043 (2004)

3. Sharbrough, F., Chatrian, G., Lesser, R., Luders, H., Nuwer, M., Picton, T.: American electroencephalographic society guidelines for standard electrode position nomenclature. Electroencephalogr. Clin. Neurophysiol. **8**, 200–202 (1991)

Chapter 6
Advanced Usage

The previous chapters provided an overview of the BCI2000 environment, and provided tutorials for commonly-used functionality. BCI2000 also provides a variety of advanced techniques for controlling and analyzing experiments. These will be covered in this chapter.

6.1 Matlab MEX Interface

6.1.1 Introduction

The BCI2000 distribution includes Matlab MEX files for manipulating BCI2000 data files, and for other tasks. MEX files allow the execution of externally compiled code from within Matlab. Because Matlab code is interpreted at runtime, and MEX files contain binary code that has been compiled prior to use, MEX files are generally faster than equivalent Matlab code. BCI2000 MEX allow for convenient access to BCI2000 data files or functions directly from Matlab.

6.1.2 Using BCI2000 MEX Files

6.1.2.1 Microsoft Windows 32

The BCI2000 binary distribution comes with pre-compiled MEX files for 32-bit Microsoft Windows platforms. Add BCI2000/tools/mex to your Matlab path to use these files, or copy them to the directory that holds your Matlab analysis scripts.

6.1.2.2 Other Platforms

BCI2000 comes with pre-built MEX files for a number of other platforms that include 64-bit Windows, Mac OS X, and Linux 32-bit as well as 64-bit. Support for

G. Schalk, J. Mellinger, *A Practical Guide to Brain–Computer Interfacing with BCI2000*,
© Springer-Verlag London Limited 2010

these MEX files is experimental, and they may or may not work on your particular configuration.

6.1.3 Building BCI2000 MEX Files

Typically, you will use pre-built binary versions of BCI2000 MEX files. If you need to build BCI2000 MEX files yourself, please follow the instructions below.

6.1.3.1 Microsoft Windows

Compiling MEX files under Windows requires the Borland C++ compiler (a free command line version is available at http://www.codegear.com/downloads/free/cppbuilder). To build MEX files,

- open a console window,
- change to the BCI2000/src/Tools/mex directory,
- execute make from there.

For Matlab versions up to 7.5 (Release 2007b), alternatively the buildmex script may be used after configuring the mex command to use the Borland compiler. For newer versions of Matlab, the mex command does not support the Borland compiler any more; still, using the make command as described above should work with newer versions of Matlab as well.

6.1.3.2 Other Platforms

For platforms other than 32-bit Windows, BCI2000 provides a Matlab script that can be used to build MEX files.

- Independently of system libraries, Matlab creates its own execution environment for MEX files. For MEX files, this means that they usually won't work unless compiled with a version of the gcc compiler that matches the one used to build Matlab itself, and linked against libraries that match the ones used by Matlab. To check, see Matlab's list of compatible compiler versions.
- After downloading the BCI2000 source code, start Matlab and make sure that the mex command is configured to use the gcc compiler. Typing help mex from the Matlab command line will guide you in configuring the mex command.
- Change Matlab's working directory to BCI2000/src/core/Tools/mex/.
- Execute buildmex from the Matlab command line.
- Add BCI2000/tools/mex to your Matlab path to use the newly built MEX files.

6.1.4 BCI2000 MEX Functions

6.1.4.1 load_bcidat

```
[ signal, states, parameters ] ...
    = load_bcidat( 'filename1', 'filename2', ... );
```

This function loads **signal**, **state**, and **parameter** data from the files whose names are given as function arguments.

Example Examples for loading multiple files:

```
files = dir( '*.dat' );
[ signal, states, parameters ] = ...
    load_bcidat( files.name );

files = ...
    struct( 'name', uigetfile( 'MultiSelect', 'on' ) );
[ signal, states, parameters ] = ...
    load_bcidat( files.name );
```

For multiple files, the number of channels, states, and signal type must be consistent across all files. By default, signal data will be in raw A/D units, and will be represented by the smallest Matlab data type that accommodates them. To obtain signal data calibrated into physical units (μV), specify '-calibrated' as an option anywhere in the argument list. (All online signal processing in BCI2000 is performed on calibrated signals.) The **states** output variable is a Matlab struct with BCI2000 state names as member names. Because there is one state value for each signal sample, the number of state values equals the first dimension of the **signal** output variable. The **parameters** output variable is a Matlab struct with BCI2000 parameter names as member names. Individual parameter values are represented as cell arrays of strings in a **Value** struct member, and additionally as numeric matrices in a **NumericValue** struct member. When there is no numeric interpretation, the corresponding matrix entry will be **NaN**. For nested matrices, no **NumericValue** field is provided. If multiple files are given, parameter values will be those contained in the first file.

This MEX file has the option to load only a range of samples in individual files. The following command will load a subset of samples defined by first and last sample index:

```
[ signal, states, parameters ] ...
    = load_bcidat( 'filename', [first last] );
```

Specifying `[0  0]` for an empty sample range makes it possible to read states and parameters from a file without reading sample data:

```
[ signal, states, parameters ] ...
    = load_bcidat( 'filename', [0 0] );
```

6.1.4.2 save_bcidat

```
save_bcidat( 'filename', signal, states, parameters );
```

This MEX function saves **signal**, **state**, and **parameter** data into the named BCI2000 file. The **signal**, **state**, and **parameter** arguments must be Matlab structs as created by the `load_bcidat`, or `convert_bciprm` MEX files. Signal data is always interpreted as raw data, i.e., it will be written into the output file unchanged.

The output file format is deduced from the output file's extension, which may be .dat, .edf, or .gdf. When no extension is recognized, the BCI2000 dat file format is used.

6.1.4.3 convert_bciprm

This MEX function converts BCI2000 parameters from Matlab struct into string representation and back.

```
parameter_lines = convert_bciprm( parameter_struct );
```

Converts a BCI2000 parameter struct (as created by `load_bcidat`) into a cell array of strings that contain valid BCI2000 parameter definition strings. When the input is a cell array rather than a Matlab struct, `convert_bciprm` converts the definition strings into a valid BCI2000 parameter struct (ignoring the **NumericValue** field if present).

```
parameter_struct = convert_bciprm( parameter_lines );
```

6.1.4.4 mem

This MEX function estimates the power spectrum using the autoregressive (AR-based) spectral estimator that is also used online in the *ARFilter*. The calling syntax is:

```
[spectrum, frequencies] = mem(signal, parms);
```

The variables **signal** and **spectrum** have dimensions channels × values; the variable **parms** is a vector of parameter values:

- model order,
- first bin center,
- last bin center,
- bin width,
- evaluations per bin,
- detrend option (optional, 0: none, 1: mean, 2: linear; defaults to none),
- sampling frequency (optional, defaults to 1).

6.2 Operator Scripting

Operator scripts automate actions that otherwise would be performed by the user, e.g., starting or suspending system operation. Scripts may be contained in script files, or given immediately in the operator module's preferences dialog. There is also an option to specify scripts from the command line when starting the operator module. These possibilities are described in more detail below.

6.2.1 Events

Execution of particular scripts can be triggered by different events that occur during various stages of BCI2000 system operation:

- **OnConnect:** This event is triggered at startup, i.e., as soon as all modules are connected to the operator module.
- **OnSetConfig:** This event is triggered each time a set of parameters is applied to the system. This happens when the user clicks the **SetConfig** button. Execution of the **SETCONFIG** scripting command also triggers this event.
- **OnStart, OnResume:** These events correspond to the **Start/Resume** button. One of these events is also triggered when the **Running** state variable is set to 1 from a script. Whether **OnStart** or **OnResume** is triggered depends on whether the system has been running before with the current set of parameters.
- **OnSuspend:** This event is triggered when the system goes from running into suspended mode. This happens whenever the **Running** state variable changes from 1 to 0. This may happen when the user clicks **Suspend,** when the application module switches the system into suspended mode, or when a script sets the **Running** state variable to 0.
- **OnExit:** This event is triggered when the operator module quits. Execution of the **QUIT** scripting command also triggers this event.

6.2.2 Scripting Commands

BCI2000 scripts consist of sequences of the scripting commands that are terminated with either a DOS line ending sequence or a semicolon (;). When the semicolon is used to terminate commands, a line may contain multiple commands. Scripts are case-sensitive, and commands must be spelled uppercase as shown in the list below.

In the *Operator module*'s preferences dialog, scripts may be entered for each of the events listed above. Scripts may be specified as paths to script files, or as immediate one-line scripts. Entries that start with a minus sign (-) are treated as one-line scripts, which may contain multiple commands separated with semicolons.

Scripts may also be specified from the command line used to start up the operator module. There, event names are followed with the content of the respective preference entry, enclosed in double quotes ("..."). The list of currently supported scripting commands is given below:

- LOAD PARAMETERFILE <file>
  Loads a parameter file specified by its path and name. Relative paths are interpreted relative to the operator module's working directory at startup. Usually, this matches the executable's location in the prog directory. The parameter file name must not contain white space. Thus, please use HTML-type encoding for white space characters, such as Documents%20and%20Settings when referring to the Documents and Settings folder.
- SETCONFIG
  Applies current parameters to the system. Corresponds to the *SetConfig* button.
- INSERT STATE <name> <bit width> <initial value>
  Adds a state variable to the system. State variables are defined by name, bit width, and initial value. This command may not be used after system initialization has completed, i.e., its use is restricted to the **OnConnect** event.
- SET STATE <name> <value>
  Sets the named state variable to the specified integer value. Setting the Running state to 1 will start system operation; setting it to 0 will suspend the system.
- QUIT
  Quits the operator module after terminating all BCI2000 modules.
- SYSTEM <command line>
  Executes a single-line shell command.

6.2.3 Examples

Example 1 To add a state variable called **Artifact**, and to set it using the operator's function buttons, do this:

1. Enter the following line under *After All Modules Connected* in the operator's preferences dialog (note the minus sign):
   -INSERT STATE Artifact 1 0

2. Under *Function Buttons,* enter `Set Artifact` as the name of button 1, and as its command, enter (note there is no minus sign): `SET STATE Artifact 1`
3. Enter `Clear Artifact` as the name of button 2, and as its command enter `SET STATE Artifact 0`

Example 2 The following example shows how to specify script commands from the command line. It fully automates BCI2000 operation by loading a parameter file, applying parameters, starting the system once the parameters are applied, and quitting the system once the run is over. For better readability, the example is broken across lines; for execution, the command needs to be entered as a single line.

```
operat.exe
      --OnConnect   "-LOAD PARAMETERFILE
                    ..\parms\examples\CursorTask\
                    SignalGenerator.prm; SETCONFIG"
      --OnSetConfig "-SET STATE Running 1"
      --OnSuspend   "-QUIT"
```

6.3 Command Line Options

6.3.1 Operator Options

The operator module makes it possible to specify scripts from the command line. Command line options correspond to script entries in the operator's preferences dialog. The following options exist: `--OnConnect`, `--OnExit`, `--OnSetConfig`, `--OnSuspend`, `--OnResume`, and `--OnStart`.

After each option, a white space is expected, followed with a double-quote enclosed string. Examples are:

- `--OnConnect "C:\scripts\onconnect.bciscript"`
- `--OnConnect "-LOAD PARAMETERFILE ..\parms\myparms.prm"`

6.3.2 Core Module Options

Using command line options at the startup of BCI2000 modules, it is possible to run different BCI2000 modules on different machines on a network, change the data format for brain signal recordings, switch on debugging messages, or automate BCI2000 operation.

Core Modules, i.e., the **Data Acquisition**, **Signal Processing**, and **Application Module**, share the same command line syntax:

```
<ModuleName> <operator IP>:<operator port>
--<option1>-<value1>
--<option2>-<value2> ...
```

All arguments are optional.

At startup, each core module connects to the operator module. If no IP address is specified on the command line, the connection is opened to the local machine using 127.0.0.1 as an IP address. When no port is given, each module uses a specified default port. Normally, there is no need to change port numbers.

Any number of options may be given, starting with a double minus sign. As indicated above, option names and values are combined with a single minus sign, to form a continuous character string. Each option is translated into a BCI2000 parameter using its name as a parameter name, and its value as a parameter value. When a parameter with the given name already exists, its value will be changed from the default to match the value given on the command line. When no parameter with that name exists, it is added to the *System* parameter section. Parameter values must not contain white spaces on the command line; however, white space may be encoded in HTTP fashion, e.g., using '%20' as a replacement for a single space character.

In addition, there are a few miscellaneous options:

- Specifying --version will display version information, and then quit.
- The *FileFormat* option will switch between different file formats used for data recording. The output file format is determined at module startup, and cannot be changed by modifying the **FileFormat** parameter from the operator module's parameter dialog.
- The *Debug* option will make a module send Debug Messages that appear in the operator module's log window.
- The *LogKeyboard*, *LogMouse*, and *LogJoystick* options will enable recording of key presses, mouse, and joystick movements, respectively.

6.3.3 Data File Formats

This section describes which data formats are available for output and how to select one of them.

At runtime, selection of a file format is achieved by specifying a command line parameter to the source module, as in

```
gUSBampSource --FileFormat=Null
```

or

```
SignalGenerator --FileFormat=GDF
```

Typically, source modules are launched from batch files contained in the "batch" directory, and their command line parameters are specified there. Internally, support for various output formats is provided by **File Writer** classes (pieces of software) implementing the interface defined by the `GenericFileWriter` class. Thus, a programmer can add support for a new output format by deriving a new class from `GenericFileWriter`, and adding it to existing source modules.

On the command line, the value that follows the last minus character, appended with `FileWriter`, will be matched against the names of all `Generic-FileWriter` descendants present in the source module. In the first example, the `NullFileWriter` class will be used for data output (which will not produce any output file), and the `GDFFileWriter` class will be used in the second example.

BCI2000 File Format (`--FileFormat=BCI2000`) Parameters, BCI2000 state variables, and brain signal data will be written into a BCI2000 data file in the BCI2000 data format. This is also the default if no file format is specified explicitly.

Null File Format (`--FileFormat=Null`) No information is recorded. Individual filters may still write log files to the directory defined by the **DataDirectory**, **SubjectName**, and **SubjectSession** parameters.

EDF File Format (`--FileFormat=EDF`) EDF (*European Data Format*) ([1], http://www.edfplus.info/specs/edf.html) is a standard for biosignal data that is particularly popular in the area of sleep research. EDF is limited to 16-bit data. BCI2000 state variables will be mapped to additional signal channels. BCI2000 parameters cannot be represented in EDF format. Setting the **SaveAdditionalParameterFile** parameter to 1 will save a separate BCI2000 parameter file along with the EDF data file.

GDF File Format `--FileFormat=GDF` GDF is a recently designed general data format for biosignals ([3], http://biosig.sf.net). Building on EDF, GDF allows for arbitrary numeric data types, introduces an event table, and provides standardized encoding of events. At present, BCI2000 supports Version 2.10 of the GDF format specification. BCI2000 does not prescribe the meaning of its state variables. GDF, on the other hand, associates a fixed set of events with certain numeric codes. Thus, a general mapping of BCI2000 states onto GDF events is not possible. Instead, GDF events are created via a user-defined set of mapping rules in the **EventCode** parameter, which also has a set of rules predefined for the most important cases. Besides GDF events, BCI2000 state variables will also be mapped to additional signal channels the same way as for EDF. Since version 2.0, GDF provides additional header space for metadata which is organized as a sequence of tag/length/value structures. BCI2000 uses such a structure – tagged with a "BCI2000" tag – to store parameters. Within the GDF data field, the same human readable format is used that applies to BCI2000 parameter files and .dat file headers.

6.4 AppConnector

6.4.1 Introduction

The BCI2000 external application (i.e., AppConnector) interface provides a bi-directional link to exchange information with external processes running on the same machine, or on a different machine over a local network. Via the external application interface, read/write access to BCI2000 state vector information and to the control signal is possible. To list examples of possible uses, external applications may read the **ResultCode** state to access the classification result, set the **TargetCode** state to control the user's task, or get access to the control signal that is calculated by **SignalProcessing** so as to control an external output device (such as a robotic arm or a web browser). Multiple instances of BCI2000 running on separate machines may share sequencing and control signal information, allowing for interactive applications such as games.

6.4.2 Scope

The scope of this interface is to provide access to internal BCI2000 information for cases in which the generation of a full-fledged BCI2000 module is impractical. Such a case might be the control of external applications that practically do not allow full incorporation into the BCI2000 framework (such as the Dasher system [2] for efficient low-bandwidth spelling).

This interface is not intended to replace the existing BCI2000 framework for BCI2000 communication. The advantages of writing modules that are fully integrated into the BCI2000 framework are that their configuration is achieved through the same interface as other BCI2000 configurations, that this configuration is stored in the data file along with all other system parameters, and that the state of the module at any given time is encoded in event markers that are also stored in the data file.

In contrast, control of an external device using the AppConnector interface implies that the configuration of the external device has to be done outside of BCI2000, that this corresponding configuration is not stored along with the data file, and that the internal state of the output device is not automatically saved together with the brain signals (although it is possible to introduce your own state variables for this purpose using the operator module's INSERT STATE scripting command). Having no configuration and state information present in the data file will make it more difficult to reconstruct what exactly was going on during an experimental session. It is thus important to keep this in mind when using this possibility.

6.4.3 Design

The design of the external application interface aims at simplicity, and at minimal interference with the timing of the signal flow through the BCI2000 system. With this in mind, we chose a connection-less, UDP-based transmission protocol rather than one based on TCP. This comes at the cost of a possible loss, or reordering of protocol messages. To keep the probability for such losses as low as possible, and their consequences as local as possible, messages have been designed to be short, self-contained, and redundantly encoded in a human readable fashion.

The connectionless nature of UDP implies that there is no server or client in the asymmetric sense that applies for TCP connections. Rather, processes write to local or remote UDP ports, and read from local UDP ports, whenever applicable. Thus, for bi-directional communication between machine A running BCI2000 and machine B running the external application, there will be two UDP ports involved:

- a port on machine B to which BCI2000 sends its messages for use in the external application, and
- a port on machine A to which the external application sends its messages for use in BCI2000.

In most cases, both BCI2000 and the external application will run on the same machine, i.e., A and B will refer to the same machine, and both ports will be local. Still, they are distinct ports.

For communication that involves a large number of network nodes, or unreliable connections, we suggest using local UDP communication, in conjunction with locally executed TCP/IP server processes that forward messages to a TCP connection between the two remote machines.

6.4.4 Description

For each block of data processed by the BCI2000 system, two types of information are sent out and may be received from the external application interface:

- the BCI2000 internal state as defined by the values of all BCI2000 states, and
- the BCI2000 control signal.

Sending data occurs immediately after the task filter of the application module processes the data; receiving occurs immediately before the task filter. This ensures that changes resulting from user choices are sent out immediately, and that received information will immediately be available to the task filter. IP addresses and ports are user-configurable. Sending and receiving may not use the same address and port.

6.4.5 Protocol

Each message in the protocol consists of a name and a value that are separated by a white space, and that are terminated with a single newline ('\n'==0x0a) character. Names may identify

- BCI2000 states by name – then followed by an integer value in decimal ASCII representation;
- Signal elements in the form Signal(<channel>,<element>) – then followed by a float value in decimal ASCII representation. Channel and element indices are given in zero-based form.

6.4.6 Examples

```
Running 0\n
ResultCode 2\n
Signal(1,0) 1e-8\n
```

The lines above show examples for valid AppConnector transmissions. If the first line was sent to BCI2000, it would switch BCI2000 into the suspended state. It is important to note that AppConnector communication is not possible during the suspended state.

The meaning of control signals depends on the used application module. For the BCI2000 cursor task, there are up to three control signal channels (channel indices 0, 1, and 2), each of which consists of one value (i.e., element). These three channels correspond to cursor velocity in X, Y, and Z direction, respectively. For example,

```
Signal(1,0) 1e-2\n
```

would indicate a value of 1e–2 for vertical cursor velocity.

6.4.7 Parameterization from Within BCI2000

BCI2000 reads AppConnector protocol lines from a local IP socket specified by the **ConnectorInputAddress** parameter, and writes such lines to the socket specified by the **ConnectorOutputAddress** parameter. Sockets are specified by an address/port combination. Addresses may be host names, or numerical IP addresses. Address and port are separated by a colon, such as in localhost:5000 or 134.2.103.151:20321.

Incoming protocol lines are filtered by name using a list of allowed names, which are defined in the **ConnectorInputFilter** parameter. To allow signal messages, allowed signal elements must be specified including their indices. To allow all names, enter an asterisk (*) as the only list entry.

6.4.8 Examples

6.4.8.1 BCI2000 Example Code

You can find simple AppConnector examples in the `BCI2000/src/contrib/AppConnectorApplications/` directory. For example, in `ParallelSwitch` there is a simple program that uses AppConnector information from BCI2000 to control the state of the parallel port. As another example, in `AppConnectorExample` you will find a GUI application that interacts with BCI2000 states.

The following listing shows a simple C++ example that reads BCI2000 AppConnector messages using the BCI2000 sockstream class:

```cpp
#include <iostream>
#include "SockStream.h"

using namespace std;

int main( int argc, char** argv )
{
  const char* address = "localhost:5000";
  if( argc > 1 )
    address = argv[ 1 ];

  receiving_udpsocket socket( address );
  sockstream connection( socket );
  string line;
  // Print each line of BCI2000 input to stdout.
  while( getline( connection, line ) )
    cout << line << endl;

  return 0;
}
```

Note: This example program uses the BCI2000 socket stream utility classes contained in `src/shared/utils/SockStream.cpp`. You will need to add this file to your project to build the example program.

6.4.8.2 An External Application Reading Information from BCI2000, Running Locally

- Set the **ConnectorOutputAddress** parameter to a local address above 1,024, such as `localhost:5000`.
- In the external application, create a UDP socket and bind it to BCI2000's output port, i.e., `localhost:5000`.
- Read from that socket as you would from a TCP socket.

6.4.8.3 An External Application Reading Information from BCI2000, Running on a Remote Machine

- Set the **ConnectorOutputAddress** parameter to a remote address with a port above 1,024, such as 134.2.102.151:20321.
- In the external program, create a UDP socket, and bind it to the remote machine's external address, i.e., 134.2.102.151:20321 rather than localhost: 20321.
- Read from that socket as you would from a TCP socket.

6.4.8.4 An External Application Sending Information to BCI2000, Running Locally

- Set the **ConnectorInputAddress** parameter to a local address with a port above 1,024, such as localhost:5001.
- Set the **ConnectorInputFilter** to * (a single asterisk).
- In the external application, create a UDP socket and bind it to BCI2000's input port, i.e., localhost:5001.
- Write valid AppConnector lines to that socket at any time.

6.4.8.5 An External Application Sending Information to BCI2000, Running on a Remote Machine

- Set the **ConnectorInputAddress** parameter to the local machine's external address, and a port above 1,024, such as bci2000machine.yourdomain. org:20320.
- In the external program, create a UDP socket and bind it to the BCI2000 machine's external address, i.e., bci2000machine.yourdomain.org: 20320.
- Write valid AppConnector lines to that socket at any time.

6.5 Expression Filter

BCI2000 provides a powerful mechanism to manipulate and modify data processing in the filter chain without the need for recompilation. This allows signal processing results to be combined with or added to BCI2000 state values. For example, it is possible to replace brain control of a cursor with joystick control simply by entering the appropriate values into the Expression Filter matrix (this example assumes that joystick logging is enabled as described in Sect. 6.3.2):

```
JoystickXpos
JoystickYpos
```

The Expression Filter reference in Sect. 10.7.4 contains detailed information on using the Expression Filter, including several examples.

References

1. Kemp, B., Värri, A., Rosa, A.C., Nielsen, K.D., Gade, J.: A simple format for exchange of digitized polygraphic recordings. Electroencephalogr. Clin. Neurophysiol. **82**(5), 391–393 (1992)
2. MacKay, D.: Dasher – An efficient keyboard alternative. Adv. Clin. Neurosci. Rehabil. **3**(2), 24 (2003)
3. Schlögl, A.: GDF – A general dataformat for biosignals (2009). http://arxiv.org/abs/cs.DB/0608052

Chapter 7
Programming Reference

7.1 Building BCI2000

This section describes how to build BCI2000 V2.0, i.e., how to create the BCI2000 executables from their source code. This process may be slightly different for BCI2000 V3.0.

7.1.1 Why Build from Source

Typically, you will install BCI2000 from its binary distribution, and do not need to build BCI2000, unless you:

- want to always use its most current SVN version,
- want to create your own module by modifying an existing one, or create a new one from scratch.

7.1.2 Tools Required

Building/compiling BCI2000 requires Borland's C++ Builder 6, or Borland/ CodeGear Development Studio 2007/2009. The operator GUI and application modules currently require the Borland VCL library. (BCI2000 V3.0 will also support VisualStudio and MinGW, and will be based on the cross-platform library Qt.) Certain non-GUI related parts of BCI2000 can be built using the freely available Borland C++ compiler, including the Matlab MEX files.

G. Schalk, J. Mellinger, *A Practical Guide to Brain–Computer Interfacing with BCI2000*,
© Springer-Verlag London Limited 2010

7.1.3 How to Build

Command-line builds require a working installation of Borland C++ Builder 6. This section describes how to build BCI2000 from the command line, from the Borland C++ 6 IDE, and from newer versions of the Borland C++ environment.

Building BCI2000 from the Command Line

1. Open a Windows cmd shell, e.g., by choosing Run... from the Start Menu and entering cmd.
2. Change to the BCI2000 src directory by typing cd c:\BCI2000\src, if BCI2000 is located at the root level of the C drive.
3. Type make from the command prompt.
4. If you updated or edited any of the source files, and experience linker errors or other unexpected behavior, execute make clean && make all.
5. For build versioning, execute make build rather than make or make all. This will update the build information visible in the BCI2000 "Version" parameters, and in the "About" boxes of BCI2000 GUI applications.

Building BCI2000 Using the Borland C++ Builder 6 IDE

1. Open the file BCI2000/src/BCI2000.bpg with the IDE by double-clicking it.
2. Make sure the "Project Manager" view is visible to the left (choose "Project Manager" from the *View* menu to display it).
3. Right-click the topmost project (located immediately below the line reading "BCI2000"), and choose Make all from here.
4. To get rid of possible inconsistencies from earlier builds, choose Build all from here rather than Make all from here.

Compiling BCI2000 with Borland C++ Builder (BCB) 2006/2007/2009

1. When using BCB 2006, please make sure to install all available updates from http://cc.embarcadero.com/reg/bds. The originally shipped version of BCB 2006 is defective and will not work without updating. We recommend updating to BCB 2007 or BCB 2009.
2. Open the file BCI2000/src/BCI2000.bpg from within Borland Developer Studio. This will import all existing projects into BDS project files.
3. In the project manager (at the top right), right-click the topmost project and choose Make all from here.
4. To get rid of possible inconsistencies from earlier builds, choose Build all from here rather than Make all from here.

> **Compiling BCI2000 with Borland C++ Builder (BCB) 2006/2007/2009** (continued)
>
> 5. Command-line builds cannot be done with Borland C++ Builder 2007/2009 because it lacks a project-to-makefile conversion utility. Still, the Matlab MEX files may be compiled from the command line as described in Sect. 6.1.
> 6. When using BCB 2006, you may experience *invalid property* runtime errors when running BCI2000. This may be due to imperfect import of *.dfm files. Make sure to install all available updates, then import a clean version of the BCI2000 source tree. If this doesn't help, consider updating to a newer version of the Developer Studio.

7.1.3.1 Building Contributions

From the Command Line   After building BCI2000 from the command line, the makefiles file in the BCI2000/src directory contains a full list of subprojects. Those from the contribution section are commented out with a # character in front. (BCI2000 consists of *core* components that were implemented by the BCI2000 development team, and *contributed* components that were designed by external contributors.) When you remove the # character, and do a make all or make build, the additional project will be included into the build. Likewise, you may speed up builds by commenting out parts of the core distribution that you don't need.

From the IDE   Open the contrib project group in the src/contrib directory, and build the contributions you want.

7.1.4 Starting up BCI2000

After compilation, a desired configuration of BCI2000 may be started by executing an appropriate batch file from the top level BCI2000/batch directory. Once a particular configuration of BCI2000 is started, simply load a parameter file with the same name as the batch file from the BCI2000/parms/examples directory. If you need a configuration for which no batch file exists, just modify a copy of a batch file that is close to your needs.

7.2 Writing a Custom Source Module

In this section, we walk you through the steps to write a new source module for a fictional data board. The concepts presented here will likely be similar for any data acquisition board that provides a C/C++ interface.

Data acquisition modules consist of code that provides generic BCI2000 framework functions (such as data storage), and code that implements access to specific data acquisition hardware. When implementing a source module, you only need to be concerned with the latter. Specifically, you need to implement a function that waits for and reads A/D data, and several helper functions that perform initialization and cleanup tasks. Together, these functions form a class derived from `Generic-ADC`.

7.2.1 Example Scenario

Your **Tachyon Corporation** A/D card comes with a C-style software interface declared in a header file `TachyonLib.h` that consists of three functions:

```
#define TACHYON_NO_ERROR 0
int TachyonStart( int inSamplingRate,
                  int inNumberOfChannels );
int TachyonStop( void );
int TachyonWaitForData( short** outBuffer,
                        int inCount );
```

From the library help file, you learn that `TachyonStart` configures the card and starts acquisition to some internal buffer; that `TachyonStop` stops acquisition to the buffer, and that `TachyonWaitForData` will block execution until the specified amount of data has been acquired, and that it will return a pointer to a buffer containing the data in its first argument.

Each of the functions will return zero if everything went well, otherwise some error value will be returned. Luckily, **Tachyon Corporation** gives you just what you need for a BCI2000 source module, so implementing the ADC class is quite straightforward.

As a remark, the .lib file provided with your device may be in a format that is not suitable for the Borland compiler. In this case, use the `implib` tool that comes with Borland to create a Borland-compatible import library (.lib file) for your device's .dll file.

7.2.2 Writing the ADC Header File

In your class' header file, `TachyonADC.h`, you write:

```
#ifndef TACHYON_ADC_H
#define TACHYON_ADC_H

#include "GenericADC.h"

class TachyonADC : public GenericADC
```

```
(continued)
{
 public:
   TachyonADC();
   ~TachyonADC();

   void Preflight( const SignalProperties&,
                   SignalProperties& ) const;
   void Initialize( const SignalProperties&,
                    const SignalProperties& );
   void Process( const GenericSignal&,
                 GenericSignal& );
   void Halt();

 private:
   int  mSourceCh,
        mSampleBlockSize,
        mSamplingRate;
};
#endif // TACHYON_ADC_H
```

7.2.3 ADC Implementation

In the .cpp file, you will need some #includes, and a filter registration:

```
#include "TachyonADC.h"
#include "Tachyon/TachyonLib.h"
#include "BCIError.h"

using namespace std;

RegisterFilter( TachyonADC, 1 );
```

From the constructor, you request parameters and states that your ADC needs; from the destructor, you call Halt to make sure that your board stops acquiring data whenever your class instance gets destructed:

```
TachyonADC::TachyonADC()
: mSourceCh( 0 ),
  mSampleBlockSize( 0 ),
  mSamplingRate( 0 )
```

```
(continued)
{
  BEGIN_PARAMETER_DEFINITIONS
    "Source int SourceCh=         64 64 1 128 "
        "// number of digitized channels",
    "Source int SampleBlockSize= 16 5 1 128 "
        "// number of samples transmitted at a time",
    "Source int SamplingRate=     128 128 1 4000 "
        "// the sample rate",
  END_PARAMETER_DEFINITIONS
}

TachyonADC::~TachyonADC()
{
  Halt();
}
```

7.2.4 ADC Initialization

Your `Preflight` function will check whether the board works with the parameters requested, and communicate the dimensions of its output signal:

```
void TachyonADC::Preflight(
                     const SignalProperties&,
                     SignalProperties& outputProperties )
                     const
{
  if( TACHYON_NO_ERROR !=
        TachyonStart( Parameter( "SamplingRate" ),
                      Parameter( "SourceCh" )
                    )
    )
    bcierr << "SamplingRate and/or SourceCh parameters"
           << " are not compatible"
           << " with the A/D card"
           << endl;
  TachyonStop();
  outputProperties = SignalProperties(
                        Parameter( "SourceCh" ),
                        Parameter( "SampleBlockSize" ),
                        SignalType::int16 );
}
```

Here, the last argument of the `SignalProperties` constructor determines not only the type of the signal propagated to the BCI2000 filters but also the format

of the dat file written by the source module. You might want to write Signal-
Type::int32 or SignalType::float32 instead if your data acquisition
hardware acquires data in one of those formats.

The actual Initialize function will only be called if Preflight did not
report any errors. Thus, you may skip any further checks, and write:

```
void TachyonADC::Initialize(
                    const SignalProperties&,
                    const SignalProperties& )
{
  mSourceCh = Parameter( "SourceCh" );
  mSampleBlockSize = Parameter( "SampleBlockSize" );
  mSamplingRate = Parameter( "SamplingRate" );
  TachyonStart( mSamplingRate, mSourceCh );
}
```

Balancing the TachyonStart call in the Initialize function, your Halt
function should stop all asynchronous activity that your ADC code initiates:

```
void TachyonADC::Halt()
{
  TachyonStop();
}
```

7.2.5 Data Acquisition

Note that the Process() function may not return unless the output signal is filled
with data, so it is crucial that TachyonWaitForData is a blocking function.
(If your card does not provide such a function, and you need to poll for data, don't
forget to call Sleep(0) or Sleep(1) inside your polling loop to avoid
tying up the CPU.)

```
void TachyonADC::Process( const GenericSignal&,
                          GenericSignal& outputSignal )
{
  int valuesToRead = mSampleBlockSize * mSourceCh;
  short* buffer;
  if( TACHYON_NO_ERROR == TachyonWaitForData(
                          &buffer,
                          valuesToRead )
    )
```

```
(continued)
  {
    int i = 0;
    for( int ch = 0; ch < mSourceCh; ++ch )
      for( int s = 0; s < mSampleBlockSize; ++s )
        outputSignal( ch, s ) = buffer[ i++ ];
  }
  else
    bcierr << "Error reading data" << endl;
}
```

7.2.6 Adding the SourceFilter

Most measurement equipment comes with hardware filters that allow you to filter out line noise. For equipment that does not offer such an option, consider adding the SourceFilter to your data acquisition module as described in Sect. 10.3.4.

7.2.7 Finished

You are done! Use your TachyonADC.cpp to replace the GenericADC descendant in an existing source module, add the TachyonADC.lib shipped with your card (or the one you created using implib) to the project, compile, and link.

7.3 Writing a Custom Signal Processing Module

The tutorial presented in this section describes how to derive a new filter class from GenericFilter, how to check preconditions, initialize your filter, and process data. It will also show you how to visualize the output signal of the filter and present it to the user.

7.3.1 A Simple Low-Pass Filter

The goal in this tutorial is to implement a low-pass filter with a time constant T (given in units of a sample's duration), a sequence $S_{in,t}$ as input and a sequence $S_{out,t}$ as output (where t is a sample index proportional to time), and obeying:

$$S_{out,0} = \left(1 - e^{-1/T}\right) S_{in,0}$$
$$S_{out,t} = e^{-1/T} S_{out,t-1} + \left(1 - e^{-1/T}\right) S_{in,t}$$

7.3.2 *The Filter Skeleton*

The resulting filter class is to be called `LPFilter`. We create two new files, `LPFilter.h`, and `LPFilter.cpp`, and put a minimal filter declaration into `LPFilter.h`:

```
#ifndef LP_FILTER_H
#define LP_FILTER_H

#include "GenericFilter.h"

class LPFilter : public GenericFilter
{
 public:
   LPFilter();
   ~LPFilter();

   void Preflight( const SignalProperties&,
                   SignalProperties& ) const;
   void Initialize( const SignalProperties&,
                    const SignalProperties& );
   void Process( const GenericSignal&,
                 GenericSignal& );
};
#endif // LP_FILTER_H
```

Into `LPFilter.cpp` we put the lines:

```
#include "PCHIncludes.h"
#pragma hdrstop

#include "LPFilter.h"

#include "MeasurementUnits.h"
#include "BCIError.h"
#include <vector>
#include <cmath>

using namespace std;
```

7.3.3 *The* `Process` *Function*

When implementing a filter, a good strategy is to begin with the `Process` function, and to consider the remaining class member functions mere helpers, mainly determined by the code of Process. First, we convert the filter's algorithm into the Process code, introduce member variables *ad hoc*, ignore possible error conditions, and disregard efficiency considerations:

```
void LPFilter::Process( const GenericSignal& input,
                        GenericSignal& output )
{
  for( int ch = 0; ch < input.Channels(); ++ch )
  {
    for( int s = 0; s < input.Elements(); ++s )
    {
      mPreviousOutput[ ch ] *= mDecayFactor;
      mPreviousOutput[ ch ] +=
          input( ch, s ) * ( 1.0 - mDecayFactor );
      output( ch, s ) = mPreviousOutput[ ch ];
    }
  }
}
```

7.3.4 The *Initialize* Member Function

As you will notice when comparing Process to the equations above, we introduced member variables that represent these sub-expressions:

$$\text{mPreviousOutput}[\,] = S_{out,t-1}$$

$$\text{mDecayFactor} = e^{-1/T}$$

We introduce these members into the class declaration, adding the following lines after the Process declaration:

```
private:
  float            mDecayFactor;
  std::vector<float> mPreviousOutput;
```

The next step is to initialize these member variables by introducing filter parameters as needed. This is done in the Initialize member function: we write it down without considering possible error conditions:

```
void LPFilter::Initialize(
      const SignalProperties& inputProperties,
      const SignalProperties& outputProperties )
```

```
(continued)
{
  // This will initialize all elements with 0
  mPreviousOutput.clear();
  mPreviousOutput.resize( inputProperties.Channels(),
                          0 );
  float timeConstant = Parameter( "LPTimeConstant" );
  mDecayFactor = ::exp( -1.0 / timeConstant );
}
```

Now this version is quite inconvenient for a user going to configure our filter, because the time constant is given in units of a sample's duration, resulting in a need to re-configure each time the sampling rate is changed. A better idea is to let the user choose whether to give the time constant in seconds or in sample blocks.

To achieve this, BCI2000 comes with a utility class MeasurementUnits that has a member ReadAsTime() that returns values in units of sample blocks (i.e., the natural time unit in a BCI2000 system). Writing a number followed by an s will allow the user to specify a time value in seconds; writing a number without the s will be interpreted as sample blocks. Thus, our user friendly version of Initialize reads:

```
void LPFilter::Initialize( const SignalProperties&,
                           const SignalProperties& )
{
  ...
  // Get the time constant in units of a sample
  // block's duration:
  float timeConstant = MeasurementUnits::ReadAsTime(
                         Parameter( "LPTimeConstant" )
                       );
  // Convert it into units of a sample's duration:
  timeConstant *= Parameter( "SampleBlockSize" );

  mDecayFactor = ::exp( -1.0 / timeConstant );
}
```

7.3.5 The `Preflight` Function

Up to now, we have not considered any error conditions that might occur during execution of our filter code. Scanning through the Process and Initialize code, we identify a number of implicit assumptions:

1. The time constant is not zero; otherwise, a division by zero will occur.
2. The time constant is not negative; otherwise, the output signal is no longer guaranteed to be finite, and a numeric overflow may occur.

3. The output signal is assumed to hold at least as much data as the input signal contains.

The first two assumptions may be violated if a user enters an illegal value into the **LPTimeConstant** parameter. In such a case, we need to make sure that an error is reported and no code is executed that depends on these two assumptions. For the last assumption, we request an appropriate output signal from the `Preflight` function. Thus, the `Preflight` code reads:

```
void LPFilter::Preflight(
        const SignalProperties& inputProperties,
        SignalProperties& outputProperties ) const
{
    float LPTimeConstant
        = MeasurementUnits::ReadAsTime(
            Parameter( "LPTimeConstant" )
        );

    LPTimeConstant *= Parameter( "SampleBlockSize" );

    // The PreflightCondition macro will automatically
    // generate an error message if its argument
    // evaluates to false.
    // However, we need to make sure that its
    // argument is user-readable
    // -- this is why we chose a variable name that
    // matches the parameter name.
    PreflightCondition( LPTimeConstant > 0 );

    // Alternatively, we might write:
    if( LPTimeConstant <= 0 )
        bcierr << "The LPTimeConstant parameter must"
               << " be greater 0"
               << endl;

    // Request output signal properties:
    outputProperties = inputProperties;
}
```

7.3.6 Constructor and Destructor

Because we do not explicitly acquire resources, and do not perform asynchronous operations, there is nothing to be done inside the `LPFilter` destructor. Our constructor will contain initializers for the members we declared, and a BCI2000 parameter definition for **LPTimeConstant**. Specifying the empty string (%) for both low and high range tells the framework not to perform an automatic range check on that parameter.

```
LPFilter::LPFilter()
: mDecayFactor( 0 ),
  mPreviousOutput( 0 )
{
  BEGIN_PARAMETER_DEFINITIONS
    "Filtering float LPTimeConstant= 16s"
    " 16s % % // time constant for the low pass"
      " filter in blocks or seconds",
  END_PARAMETER_DEFINITIONS
}

LPFilter::~LPFilter()
{
}
```

7.3.7 Filter Instantiation

To have our filter instantiated in a signal processing module, we add a line that contains the `Filter` statement to the module's `PipeDefinition.cpp`. This statement expects a string parameter which is used to determine the filter's position in the filter chain. If we want to use the filter in the AR Signal Processing module, and place it after the `SpatialFilter`, we add:

```
#include "LPFilter.h"
...
Filter( LPFilter, 2.B1 );
```

to the file `SignalProcessing/AR/PipeDefinition.cpp`. Now, if we compile and link the AR Signal Processing module, we get an "unresolved external" linker error that reminds us to add our `LPFilter.cpp` to that module's project.

7.3.8 Visualizing Filter Output

Once our filter has been added to the filter chain, the BCI2000 framework will automatically create a parameter **VisualizeLPFilter** that is accessible under **Visualize → Processing Stages** in the operator module's configuration dialog. This parameter allows the user to view the LPfilter's output signal in a visualization window. In most cases, this visualization approach is sufficient. For the sake of this tutorial, however, we will disable automatic visualization, and implement our own signal visualization.

To disable automatic visualization, we override the `GenericFilter::AllowsVisualization()` member function to return `false`. In addition, to present the LPFilter's output signal in an operator window, we introduce a member of type `GenericVisualization` into our filter class:

```
#include "GenericVisualization.h"
...
class LPFilter : public GenericFilter
{
  public:
...
    virtual bool AllowsVisualization()
      const { return false; }

  private:
...
    GenericVisualization  mSignalVis;
};
...
```

`GenericVisualization`'s constructor takes a string-valued visualization ID as a parameter; we need to get a unique ID in order to get our data routed to the correct operator window. Given the circumstances, the string "LPFLT" appears unique enough, so we change the `LPFilter` constructor to read:

```
LPFilter::LPFilter()
: mDecayFactor( 0 ),
  mPreviousOutput( 0 ),
  mSignalVis( "LPFLT" )
{
  BEGIN_PARAMETER_DEFINITIONS
    "Filtering float LPTimeConstant= 16s"
      " 16s % % // time constant for the"
        " low pass filter in blocks or seconds",
    "Visualize int VisualizeLowPass= 1"
      " 1 0 1 // visualize low pass"
        " output signal (0=no, 1=yes)",
  END_PARAMETER_DEFINITIONS
}
```

In `Initialize`, we add

```
mSignalVis.Send(
  CfgID::WindowTitle, "Low Pass" );
mSignalVis.Send(
  CfgID::GraphType, CfgID::Polyline );
mSignalVis.Send(
  CfgID::NumSamples,
  2 * Parameter( "SamplingRate" ) );
```

Finally, to update the display in regular intervals, we add the following at the end of `Process`:

```
if( Parameter( "VisualizeLowPass" ) == 1 )
  mSignalVis.Send( output );
```

We might also send data to the already existing task log memo window, adding another member:

```
GenericVisualization  mTaskLogVis;
```

initializing it with:

```
LPFilter::LPFilter()
: ...
  mTaskLogVis( SourceID::TaskLog )
{
  ...
}
```

and, from inside `Process`, writing some text to it as in:

```
if( output( 0, 0 ) > 10 )
{
  mTaskLogVis << "LPFilter: (0,0) entry of"
              << " output exceeds 10 and is "
              << output( 0, 0 )
              << endl;
}
```

7.4 Writing a Custom Matlab Filter

7.4.1 Online Algorithm Verification

In the field of BCI signal processing research, novel methods are often proposed and tested on the basis of existing data. While novel methods can clearly improve performance, it is important to keep in mind that timely feedback of brain signal classification is an essential element of a BCI. Thus, any proposed BCI signal processing method should be verified with respect to its viability and usefulness in a true online setting, in which feedback is provided to the subject in real time.

Such novel methods are usually first developed and tested offline. BCI2000 facilitates the transformation of an existing offline analysis method into a functional online system. To do this, it provides a convenient and simple programming interface. Additionally, a signal processing component may be implemented as a set of Matlab scripts. It is still easy to underestimate the effort required to transform an existing offline implementation of a signal processing algorithm into a functional online implementation. While BCI2000 tries to make the transformation as simple as possible, it cannot remove the effort required to deal with chunks of data, which implies the need for

- buffering: rather than having immediate access to a continuous data set, it may be necessary to maintain an additional data buffer;
- maintaining a consistent state between subsequent calls to the processing script (when using the Matlab interface).

7.4.2 An Example Algorithm in Matlab

In this scenario, we use a simple, straightforward BCI signal processing algorithm designed for the mu rhythm paradigm. This serves as an example to illustrate the necessary steps to modify the algorithm such that it may be used to build an online system in BCI2000.

In the example, signal processing consists of IIR bandpass filtering, followed with envelope computation, and linear classification. A typical Matlab implementation of that algorithm might consist of about ten lines:

```
function class_result = classify( data, band, classifier );

% Use as
%    class_result = classify( data, band, classifier )
%
% This function takes raw data as a [channels x samples]
% vector in the 'data' input variable.
%
% Then, it computes bandpower features for the band specified
% in the 'band' input variable, which is a number between 0
% and 0.5, specifying center frequency in terms of the
% sampling rate.
%
```

```
(continued)
% As a last step, it applies the 'classifier' matrix to the
% features in order to obtain a single classification result
% for each sample. The 'classifier' vector specifies a
% classification weight for each processed channel.
%
% The result is a single classification result for each
% sample.
%
% This requires the Matlab signal processing toolbox.

% Design bandpass filters and apply them to the data matrix.
% The filtered data will contain bandpass filtered data as
% channels.
[n, Wn] = buttord(band*[0.9 1.1]/2, band*[0.7 1.4]/2,1,60);
[b, a] = butter(n, Wn);
processed_data = filter(b, a, data);

% For demodulation, rectify and apply a low pass.
[b, a] = butter(1, band/4);
processed_data = filter(b, a, abs(processed_data));

% Finally, apply the linear classifier.
class_result = processed_data * classifier;
```

Note that, to be viable in an online environment, an algorithm must operate on its signal in a causal way, i.e., it may not use future samples in order to process present samples. (A certain amount of non-causality may be possible by windowed operation on buffered data, although this will increase the effective delay between input and output data.)

Also note that the classify() function omits spatial filtering – in Matlab, this may be done easily by pre-multiplying the data with a spatial filter matrix if desired.

7.4.3 Transforming Matlab Code into BCI2000 Events

BCI2000 can execute Matlab code using the MatlabSignalProcessing module. For use with BCI2000, our Matlab code needs to be cast into a form that suits the BCI2000 filter interface. In this event-based model, which is very similar to that described earlier in this chapter for C++, portions of code are called at certain times to configure a filter component's internal state, and to act upon the signal in chunks on its way through the BCI2000 chain of filters.

7.4.3.1 Process

The most central event in the filter interface is the "Process" event. The "Process" event handler receives a signal input, will process this input according to the filter component's role in the signal processing chain, and return the result of processing in a signal output variable. It is important to understand that the "Process" handler is called separately for each chunk of data, and thus does not see the signal in its entirety. Also, the size of data blocks (chunks) is freely configurable by the user. This implies that "Process" scripts may not depend on a certain data block size, and will sometimes need to maintain their own data buffers when the algorithm in question operates on windows of data rather than continuously. In the current example, this is not the case, so we need not maintain an internal data buffer. Still, we need to maintain an internal state between calls to the "Process" handler in order to preserve the state of IIR filter delays. This will allow continuous operation on the signal with the same processing result as in the offline version of the algorithm.

The algorithm's online version in the bci_Process script will thus be:

```
function out_signal = bci_Process( in_signal )

% Process a single block of data by applying a filter to
% in_signal, and return the result in out_signal.
% Signal dimensions are ( channels x samples ).

% We use global variables to store classifier,
% filter coefficients and filter state.
global a b z lpa lpb lpz classifier;

[out_signal, z] = filter( b, a, in_signal, z );
out_signal = abs( out_signal );
[out_signal, lpz] = filter( lpb, lpa, out_signal, lpz );
out_signal = out_signal * classifier;
```

7.4.3.2 Initialize

Determination of filter coefficients is part of per-run initialization, and occurs in the "Initialize" event handler:

```
function bci_Initialize( in_signal_dims, out_signal_dims )

% Perform configuration for the bci_Process script.

% Parameters and states are global variables.
global bci_Parameters bci_States;

% We use global variables to store classifier vector,
% filter coefficients and filter state.
global a b z lpa lpb lpz classifier;

% Configure the Bandpass filter
band = str2double( bci_Parameters.Passband ) / ...
         str2double( bci_Parameters.SamplingRate );
```

```
(continued)
[n, Wn] = buttord(band*[0.9 1.1]/2, band*[0.7 1.4]/2,1,60);
[b, a] = butter(n, Wn);
z = zeros(max(length(a), length(b))-1, in_signal_dims(1));

% Configure the lowpass filter
[lpb, lpa] = butter(1, band/4);
lpz = zeros(max(length(lpa), length(lpb))-1, ...
      in_signal_dims(1));

% Configure the Classifier vector
classifier = str2double( bci_Parameters.ClassVector );
```

7.4.3.3 StartRun

In addition, we need to reset filter state at the beginning of each run, using the "StartRun" event handler:

```
function bci_StartRun

% Reset filter state at the beginning of a run.

global z lpz;
z   = zeros(size(z));
lpz = zeros(size(lpz));
```

7.4.3.4 Constructor

To complete the Matlab filter code, we need to declare the "Band" and "Classifier" parameters in a "Constructor" event handler:

```
function [ parameters, states ] = bci_Construct

% Request BCI2000 parameters by returning parameter definition
% lines as demonstrated below.

parameters = { ...
  [ 'BPClassifier float Passband= 10 10 0 % %' ...
      ' // Bandpass frequency in Hz' ] ...
  [ 'BPClassifier matrix ClassVector= 1 1 1 0 % %' ...
      ' // Linear classifier vector' ] ...
};
```

7.4.3.5 Preflight

Finally, we need to check parameters for consistency in a "Preflight" script, and declare the size of our filter's output signal:

```
function [ out_signal_dim ] = bci_Preflight( in_signal_dim )

% Check whether parameters are accessible, and whether
% parameters have values that allow for safe processing by the
% bci_Process function.
% Also, report output signal dimensions in the
% 'out_signal_dim' argument.

% Parameters and states are global variables.
global bci_Parameters bci_States;

band = str2double( bci_Parameters.Passband ) ...
         / str2double( bci_Parameters.SamplingRate );
if( band <= 0 )
  error( 'The Passband parameter must be greater zero' );
elseif( band > 0.5 / 1.4 )
  error( [ 'The Passband parameter conflicts with the ' ...
    'sampling rate' ] );
end

out_signal_dim = [ 1, size( in_signal_dim, 2 ) ];
if( in_signal_dim( 1 ) ~= ...
      size( bci_Parameters.ClassVector, 2))
  error( [ 'ClassVector length must match the input ' ...
    'signal''s number of channels' ] );
end
```

In addition to this tutorial, you can find more information about Matlab support in Sect. 10.7.5.

Chapter 8
Exercises

The previous chapters provided descriptions of different aspects of the BCI2000 system, and gave tutorials for use of two different types of brain signals. In this chapter, we include a number of exercises that build on these descriptions. Each of these exercises addresses a different component of the system. The exercises are meant to be followed in order.

Chapters 3 and 10 give an overview and detailed description of the most important components of the BCI2000 system, respectively. We will here provide a realistic exercise that will touch on the key parameters of these components. This exercise will simulate the configuration of a BCI system that implements a cursor task. Thus, all subsequent exercises assume the use of the `Cursor-Task_SignalGenerator.bat` batch file. We suggest to start with the `Cursor-Task_SignalGenerator.prm` parameter file and to change it appropriately as you go along. It is a good idea if you save the parameter file with a different name as you make changes.

To begin, start BCI2000 using the batch and parameter files listed above. You can press *Set Config* and *Start*. You will notice that you can control the cursor in the display using the mouse. The exercises in the following sections will reconfigure a number of components of the system. The cursor will follow the mouse again only when all configuration changes have been made.

8.1 Source Module

We begin with an exercise relating to signal acquisition. Thus, we will be changing parameters in the *Source* tab of the BCI2000 configuration dialog.

Q: We assume that we are recording EEG signals from eight channels (C3, Cz, C4, Cp3, Cpz, Cp4, Fpz, Oz) using a data acquisition (DAQ) device that has a sampling rate of 512 Hz. The whole system (i.e., including cursor movement) should be updated 32 times per second. The DAQ device provides floating point signals that are in units of μV. Which BCI2000 parameters need to be changed and why?

G. Schalk, J. Mellinger, *A Practical Guide to Brain–Computer Interfacing with BCI2000*, 119
© Springer-Verlag London Limited 2010

A: Because we are simulating acquisition at 512 Hz from eight channels, we need to set the parameters **SamplingRate** to 512 and **SourceCh** to 8, respectively. Because we know the names of the channels, we can enter them (delimited by spaces) in the **ChannelNames** parameter. This way, we can refer to channels by names when BCI2000 processes signals. Because the DAQ device gives floating point signals in microvolts, we need to set **SignalType** to *float32* and **SourceCh-Gain** to a series of ones (i.e., the conversion factor between the input signal from the DAQ device and μV is 1), respectively. Because there are eight channels, **SourceCh-Gain** needs to be a series of eight 1's and **SourceChOffset** needs to be a series of eight 0's. (If very precise measurements are important, it may be important to determine the offset and gain of each channel using an external procedure and then to configure these two parameters with specific values determined from that procedure.) Because we would like to update the system 32 times per second, we need to set **SampleBlockSize** to 16 (i.e., 512 samples per second divided by 32 updates per second equals 16 samples per update/block).

At this point, you can click on *Set Config* and then *Start*. You will notice a window labeled *Source Signal* with simulated EEG. The simulated EEG consists of noise and sine waves whose amplitude responds to mouse movements. Notice that the channel labels correspond to the channels you defined. You should also see the simulated EEG and the cursor movement updating smoothly. Also notice that in the *Timing* window, the two bottom-most lines (i.e., **Roundtrip** and **Stimulus**) reach up to a sizable part of the upper-most trace (i.e., **Block**). This means that it takes the system a good portion of one block to process that block and to update the display. In this respect, note that one block only corresponds to 31.25 ms $\left(\frac{16 \ (=\text{SampleBlockSize})}{512 \ \text{Hz}} \times 1,000 \ \frac{\text{ms}}{\text{s}} \right)$. As a test, you may now change the **Sample-BlockSize** to 128, set the configuration again and start. You will notice that the *Source Display* and cursor movement are much choppier (since the display is only updated four times per second ($\frac{512}{128}$)), but that now the **Roundtrip** and **Stimulus** traces in the **Timing** window only capture a very small portion of the **Block** trace. In other words, the Timing window gives a simple view of how taxed the computer is with your particular BCI2000 configuration. This is determined by the performance of your computer (in particular, CPU speed, number of processors/cores, and graphics card performance) and the complexity of the BCI2000 configuration (i.e., block size, sampling rate, number of channels, and the demands of signal processing and visual display). One common situation is that the visual display begins to consume substantial processing cycles at higher sampling rates. Thus, if the **Roundtrip** and **Stimulus** traces get closer to or even exceed the **Block** trace, you may want to increase **VisualizeSourceDecimation** in the *Visualization* tab. Set the parameter **SampleBlockSize** back to 16.

Q: In this exercise, we will use the mouse to modulate the amplitude of a sine wave at C3 and C4 at 18 Hz. This simulates a subject that can modulate the amplitude of a sensorimotor rhythm at these locations and that frequency. With our BCI, we would like to control a computer cursor using one or both of these signals. Which BCI2000 parameters need to be changed and why?

A: As described, we want to simulate a sine wave at channel C3 (channel 1) and C4 (channel 3) at 18 Hz. Thus, we will set parameters **SineChannelX** to 1, **SineChannelY** to 3, and **SineFrequency** to 18 Hz, respectively. We also make sure that **ModulateAmplitude** is checked. Set the configuration and start. Notice that the sine wave at channels 1 (C3) and 3 (C4) will respond to horizontal and vertical mouse movements, respectively.

8.2 Signal Processing Module

In the next part of the exercise, we will be extracting the simulated signals (i.e., amplitude of the sine waves) at C3 and C4 and 18 Hz in real time and will be using it for cursor control. To do this, we will have to deal with at least four components of Signal Processing, i.e., the **Spatial Filter**, the **AR Filter**, the **Linear Classifier**, and the **Normalizer**. The following exercises address these three components.

Q: Spatial filtering can substantially improve the results of feature extraction in EEG-based BCIs [3]. Thus, while not necessary for our simulated example, we will apply a **Common Average Reference (CAR)** filter. This filter subtracts the average of all other channels from a particular channel.[1] In our example, in which we expect signals at channels C3 and C4, we would thus create two output signals that each have the average of all other channels subtracted. This can be done using the **Spatial Filter**. Which configuration changes are necessary?

A: Two changes are necessary. The first configuration change ensures that all eight channels are being transmitted to the **Signal Processing** module. By default, not all channels are being submitted to **Signal Processing** to reduce processing demands. Which channels are transmitted is defined by the parameter **TransmitCh-List** in the *Source* tab. Because we would like to transmit all eight channels, this parameter needs to be set to 1 2 3 4 5 6 7 8. Thus, there are eight channels available to the **Signal Processing** module, which are identical to the eight acquired channels.

The second configuration change implements a spatial filter with the desired characteristics. As described, we would like to subtract from C3 or C4 the average of all the other channels to yield the filtered channels C3f and C4f. Thus, this can be written as `C3f=C3-(Cz+C4+Cp3+Cpz+Cp4+Fpz+Oz)/7` and `C4f=C4-(C3+Cz+Cp3+Cpz+Cp4+Fpz+Oz)/7`. Given $\frac{1}{7} = 0.14$ and reformatting for the order of the channels in the list of transmitted channels, this gives the following spatial filters:

```
C3f=1*C3-0.14*Cz-0.14*C4-0.14*Cp3-0.14*Cpz-0.14*Cp4-0.14*Fpz-0.14*Oz,
```

---

[1]Some scientists would subtract the average of *all* channels when calculating the CAR filter. It is not clear whether one version of that filtering technique is generally better than the other.

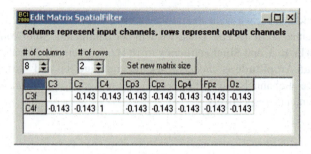

Fig. 8.1 The spatial filter matrix for this example

and

```
C4f=-0.14*C3-0.14*Cz+1*C4-0.14*Cp3-0.14*Cpz-0.14*Cp4-0.14*Fpz-0.14*Oz.
```

In other words, the spatial filter defined by the matrix shown in Fig. 8.1 takes input from all eight transmitted channels (which in this example are identical to the acquired eight channels, but do not have to be) and produces two output signals labeled C3f and C4f. Thus, all further processing is done on only two channels where the first channel represents the spatially filtered version of C3 (which was the first acquired channel and the first transmitted channel), and the second channel represents the spatially filtered version of C4 (which was the third acquired channel and the third transmitted channel). Because the input of the spatial filter is the list of transmitted channels, the number of columns in the *SpatialFilter* matrix has to equal the number of transmitted channels specified in *TransmitChList*.

Q: Once the signals are spatially filtered, we need to convert them into the frequency domain, i.e., to calculate a spectrum. Once we have derived a spectrum, we can extract the signal at 18 Hz that we know is represented in the signal. For reasons described elsewhere [2], the most commonly used spectral estimation routine in BCI2000 is the Maximum Entropy Method (MEM) method [1] rather than, for example, the Fast Fourier Transform (FFT). The parameters corresponding to the MEM method are listed in the **ARFilter** section of the *Filtering* tab. Which parameters of this section, if any, need to be changed and why?

A: No parameters in this section need to be adapted. With the settings in this parameter file, a spectrum is calculated between 0 and 30 Hz in 3 Hz bins. This spectrum is calculated based on the previous 0.5 s, which is a reasonable compromise between accuracy and group delay. The preselected model order is 16. For EEG, good values are 16–20, whereas for ECoG, good values are 20–30. In either case, the output of the **ARFilter** component consists of two channels of spectra that correspond to the spectra for C3f and C4f. Each of the two spectra consists of 11 bins that reflect the frequency estimates at 0, 3, 6, 9, 12, 15, 18, 21, 24, 27, and 30 Hz. The next task is to translate these estimated frequency amplitudes into usable outputs for cursor control.

Q: The next step is to select from these features (i.e., amplitudes at given locations and frequencies) the linear combination of features that represents the signals

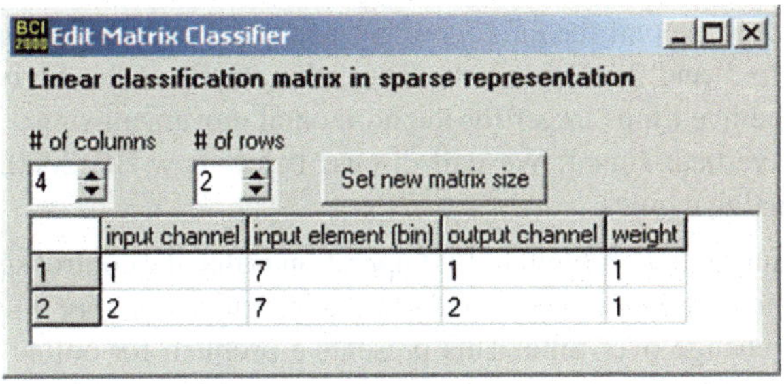

Fig. 8.2 The *Classifier* parameter for this example

of interest. In this case, we know that the signals of interest are the amplitudes at C3 and C4 at 18 Hz. We also know that activity at C3 corresponds to horizontal movement of the mouse and activity at C4 corresponds to vertical movement of the mouse. Given that we would like the same mapping for control of the cursor, which parameters need to be changed and how?

A: The relevant parameter is **Classifier** in the *Filtering* tab. As described above, the signals of interest are located at C3 (i.e., spatially filtered channel 1) and C4 (i.e., spatially filtered channel 2) at 18 Hz (i.e., bin 7 in the computed spectra). Thus, we would like to associate channel 1, bin 7, with control signal/output channel 1 (which is used as horizontal velocity in the **CursorTask**); and we would like to associate channel 2, bin 7, with control signal/output channel 2 (which is used as vertical velocity in the **CursorTask**). The full configuration is shown in Fig. 8.2. As an alternative to using channel and bin numbers, it is also possible to enter channel names and frequencies (i.e., C3f/C4f and 18 Hz, respectively).

This is a simple example for a classifier that makes use of only two features. More complex classifiers could be designed (perhaps using an automated routine offline) and then used. In such a situation, there will likely be a linear combination of features rather than just one like in this example. In this case, the resulting control signal/output channel values are linear combinations of the corresponding feature values multiplied with the weight (defined in the right-most column) for each feature.

Now it is a good time to save your parameter file. Then, when you set the configuration and start the system, you should position the mouse on the appearing targets. After several trials, the system will begin to appropriately react to the mouse movements. This is because the system has to adapt to the control signal values produced by the classifier you just designed. The function of this adaptation routine is discussed in the next exercise.

Q: For this exercise, let's change the weights of the first and second control signal. To do this, enter 5 and 2 in the first and second row and in the right-most column in the parameter **Classifier** in the *Filtering* tab, respectively. Set the configuration and start. What happens and why?

A: For the first couple of trials, the cursor automatically moves to the top right of the window. This is because we set the weight of the first and second control signal (which correspond to horizontal and vertical movement in the cursor task, respectively) to 5 and 2, respectively. Thus, the output of the **Signal Processing** module became five times larger for the horizontal movement signal and two times larger for the vertical signal. We will discuss below how BCI2000 adapts to this change in signal dynamics.

Q: One substantial problem in BCI operation is that the control signal dynamics (e.g., their minimum, maximum, mean, or standard deviation) are initially unknown and may also change over time. This presents a problem for online operation that needs to appropriately use these signals for a particular function, e.g., to control a cursor. As an example, the amplitude of a particular brain signal may change between 10 and 100 on one day, and between 100 and 300 the following day. Thus, in order to effectively use these control signals, they have to be normalized. For example, they should have zero mean and unit variance. How could this be accomplished given that we do not know much about these signals ahead of time?

A: One pragmatic and effective way to tackle this problem is to estimate the control signal mean o and standard deviation s from the control signal c derived online, and then to normalize each control signal by these values, e.g., using the following equation: $c' = (c - o)/s$. In BCI2000, these values for o (offset) and $g := 1/s$ (gain) are represented by the parameters **NormalizerOffsets** and **NormalizerGains**, respectively. Thus, the **Normalizer** component in the BCI system subtracts from each control signal, which in itself results from application of the *Classifier* matrix, the corresponding value (e.g., the second entry for the second control signal) of the **NormalizerOffsets** parameter, and multiplies the result with the corresponding value of the **NormalizerGains** parameter.

Q: Can you think of a few different approaches to estimating these offset and gain parameters from the data acquired in real time? Which approaches could you use for the following two scenarios: in scenario 1, you know something about the desired task. For example, you may know that there are four targets on the screen (like in the cursor task examples used in these exercises), and you also know what target the cursor should move towards. In scenario 2, you do not know anything about the task, e.g., like in an imaginary wheelchair application in which the user could move toward any target.

A: In scenario 1, we could simply estimate the dynamics of the control signal values for each target separately, e.g., calculate the mean and variance for the left and right target, and average them between the two targets to determine the offset (mean) and gain (1/standard deviation) for the horizontal control signal. The same procedure could be applied for the top and bottom target to determine the offset and gain for the vertical control signal. Knowing about the correct task category (i.e., target) has the advantage that we can get a good estimate of offsets and gain relatively quickly, and that it will be correct even if the target frequencies are not the same (e.g., in case the left target appears twice as often as the right target). The disadvantage is that in a real-world application (i.e., scenario 2), we will not know what the desired target is at any given point in time. In this case, we can only estimate control signal parameters from all control signal values, i.e., assuming that the control signal mean over time should be zero. This implies that we will need to use a longer period for that estimation (since it may be possible that just by chance we could move to the left five times in a row), and that we would not be able to appropriately adapt in a scenario in which one direction is always chosen more often than another. The adaptation concept in BCI2000 can actually accommodate both situations. Please see Sect. 10.6.6 for a detailed description.

Q: If you use the standard `CursorTask_SignalGenerator.bat` batch file, and `CursorTask_SignalGenerator.prm` parameter file, you can control the cursor using your mouse. This functionality was used earlier in this section to demonstrate different signal processing aspects in BCI2000. If you control the cursor using your mouse, you will notice that there is a small delay between when you move the mouse and when the cursor starts moving. What causes this delay? (Hint: this delay is not due to slow processing or screen update.)

A: In this configuration, the movement of the mouse is modulating the amplitude of a sine wave around 10 Hz. Signal processing then extracts this signal by computing a spectral analysis. Any spectral analysis has to be performed on a particular time window, which in the AR Signal Processing filter is governed by the **WindowLength** parameter. By default, this window length is set to 500 ms. In other words, a particular movement will not immediately affect the output, and will have a maximum effect on the output 500 ms later (assuming the movement lasted for the whole 500 ms). Thus, this delay is intrinsically caused by the carrying signal (a 10 Hz oscillation), and the procedure to extract that signal. It is important to realize that, on a typical machine, this delay will be dramatically longer than any delay caused by BCI2000 signal processing or feedback/screen update.

References

1. Marple, S.L.: Digital Spectral Analysis: With Applications. Prentice–Hall, Englewood Cliffs (1987)
2. McFarland, D.J., Lefkowicz, T., Wolpaw, J.R.: Design and operation of an EEG-based brain-computer interface (BCI) with digital signal processing technology. Behav. Res. Methods Instrum. Comput. **29**, 337–345 (1997)
3. McFarland, D.J., McCane, L.M., David, S.V., Wolpaw, J.R.: Spatial filter selection for EEG-based communication. Electroencephalogr. Clin. Neurophysiol. **103**(3), 386–394 (1997)

Chapter 9
Frequently Asked Questions

9.1 Timing Issues

Q: The signal traces in the source signal display, or the feedback cursor in the application display, is not updated at a constant speed, but rather appears jumping.

A: In the timing display window (enable in the *Visualize* tab if not shown), check whether block duration (the top curve) is constantly at the level indicated by the tick mark to the left. When this is not case, try the following:

- Increase the value of the **Source Decimation** parameter on the *Visualize* tab to reduce the processor load associated with the source signal display.
- Increase the value of the **SampleBlockSize** parameter to reduce the system update rate.

9.2 Startup

Q: When clicking *Set Config*, I get error messages for each of the modules such as: EEGSource: Could not make a connection to the SignalProcessing Module and Application: SignalProcessing dropped connection unexpectedly.

A: This may be caused by the *Microsoft TV\Video Connection* in Network Connections, if you have it. To remedy this:

1. Right-click on *My Network Places*,
2. Select *Properties*,
3. Right-click on *Microsoft TV\Video Connection*,
4. Select *Properties*,
5. Uncheck *Internet Protocol*,
6. Click *OK*.

Q: When I start the **gUSBamp** source module, I get the error message: Cannot find ordinal 37 in dynamic library gUSBamp.dll.

G. Schalk, J. Mellinger, *A Practical Guide to Brain–Computer Interfacing with BCI2000,*   127
© Springer-Verlag London Limited 2010

A: Most likely, your **gUSBamp** source module finds an incompatible version of the `gUSBamp.dll` driver library. Make sure that, in the directory where `gUSBampSource.exe` resides (i.e., usually the `prog` directory), or in the `\windows\system32` directory, there is a file `gUSBamp.dll` in the exact version that matches the gUSBamp driver.

Q: The BCI2000 **gUSBamp** source module does not recognize my g.USBamp amplifier. When I click *Set Config,* I get the an error message: `gUSBampADC::Preflight: Could not detect any amplifier. Make sure there is a single gUSBamp amplifier connected to your system, and switched on.`

A: Make sure that the amplifier is attached, switched on, and that it works with g.tec's demonstration program.

9.3 Feedback

Q: The first time a subject is run using the cursor feedback protocol, and during the first few trials of the first run only, the cursor moves to the top of the screen and does not respond to user input.

A: This is not so much a bug as a side effect of the way the adaptation mechanism works. Since the adaptation mechanism hasn't been trained, all it has are the initial values for offset and gain. For each subject, it will take a few trials to adapt BCI2000 to this subject's signals.

Therefore, you will typically save a subject's adapted parameters after each session, and use this as a starting point for the next session. Alternatively, you may load a subject's parameters from a previous session's `data` file by specifying the data file rather than a parameter file from the `Load Parameters...` dialog.

9.4 Replaying Recorded Data

Q: Rather than controlling BCI2000 from a brain signal, I would like to use a recorded data file as input. Is there a kind of "replay mode" in BCI2000?

A: While this could easily be done, e.g., by writing a source module that reads from a file, we have not done this, and this is for at least two reasons:

First, if the version of BCI2000 that you are using is not the exact same as the one that was used to record the data, and/or you change some system parameters, then, depending on the experimental paradigm, results may be undefined. E.g., in a 2D cursor task, what should happen if you slow down the cursor such that the cursor has not yet reached a target when in the original data file it has (i.e., and the trial is over)?

Second, scientifically it is a bad idea to "simulate" online performance by loading a single datafile, changing a few parameters, and seeing how it would have worked. If you are interested in evaluating the effect of different signal processing routines, you should write your own offline analysis routine that uses statistical analyses, together with a comprehensive body of data, or you need to run comprehensive online studies.

Of course there is nothing in the code that would prohibit you from writing your own such module, but you should note the two caveats above.

9.5 Random Sequences

Q: Is there a way to obtain the same random sequence each time BCI2000 is run? I don't care about the sequence itself, it should just be the same each time.

A: The **SignalGenerator** source module has a parameter, **RandomSeed**, that determines the pseudo random number generator's seed value. If this parameter is zero, the generator will be initialized from the system timer.

While absent from other BCI2000 modules by default, each component that uses the BCI2000 pseudo random generator will behave according to this parameter if it is present. To introduce the **RandomSeed** parameter into a module that does not provide the **RandomSeed** parameter, use --RandomSeed=10 as a command line option to that module when starting up BCI2000. Note that the parameter will have the same value across all modules (10 in the example), and it will appear on the *System* tab in the operator module's parameter dialog.

9.6 Visual Stimulation

Q: Is visual stimulation synchronized with screen refresh (vertical blanks)?

A: BCI2000 does not synchronize its video output with refresh cycles to avoid interference with its own timing. We have run comprehensive evaluations that demonstrated that with CRT monitors that are set to high refresh rates (e.g., > 100 Hz) and with appropriate computer power, the latency jitter in the BCI2000 system (i.e., from data acquisition to stimulus presentation) is less than 10 milliseconds with even smaller jitter. This timing behavior can support a large fraction of all BCI experiments. For more information about BCI2000 timing, please check out [1].

If you feel that output should be synchronized, and have access to the BCI2000 source code, adding a DirectDraw WaitForVerticalBlank() call immediately before the Task filter's call to UpdateWindow() at the end of its Process() function should help.

Q: The Stimulus Presentation program only allows for either a random sequence of stimuli, or a deterministic sequence of stimuli. In my experimental paradigm, I would like to present random stimuli, but have them interspersed by a particular resting stimulus. It looks like BCI2000 does not support that?

A: BCI2000 can present any sequence of stimuli. Referring to the example above, where stimulus 1 may be the resting stimulus and stimuli 2–9 are to be randomly presented, the following could be a valid sequence: 1 3 1 2 1 9 1 8

1 5 1 6. Such a sequence could be determined using an external procedure, and then BCI2000 could be configured using that particular sequence. Specifically, it would be possible to write a Matlab program similar to the one below.

```
num_values = 9;
r = round(rand(1, num_values)*num_values+1);
fp = fopen('fragment.prm', 'wb');
fprintf(fp, 'Application:Sequencing intlist ' ...
              'Sequence= %d', num_values*2);
for i = 1:num_values
  fprintf(fp, ' 1 %d', r(i));
end
fprintf(fp, ' 1 1 %% // test parameter\r\n');
fclose(fp);
```

Execution of that Matlab program produces a parameter file fragment fragment.prm. When loaded on top of a full parameter file, this would produce a random sequence. This process can be fully automated using BCI2000 batch files, scripting, and command line parameters (see Sect. 6.2) so that clicking on one icon could run that Matlab program, execute BCI2000, load a full parameter file, load the parameter file fragment (containing the randomized sequence) produced by Matlab, set the configuration, and start operation.

Q: I would like to control the Cursor Task in BCI2000 using a joystick or a mouse, e.g., to implement a standard center-out paradigm. How do I do that?

A: As is so often the case, there are different ways of accomplishing a particular goal. In this case, it would be possible to change the source code of the Cursor Task such that it would be controlled by, for example, a joystick. However, BCI2000 supports a solution that is much more elegant and powerful. This solution is based on the *Expression Filter* described in Sect. 10.7.4. Cursor movement in the Cursor Task is controlled by the first and second control signal produced by the Signal Processing module. Typically, this control signal is controlled by the brain signals that are result of the Linear Classifier (see Sect. 5.2.4.2 or Sect. 10.6). These resulting control signals are passed through the Expression Filter where they can be modified using an algebraic expression. This algebraic expression has access to BCI2000 states, and joystick position can easily be logged in states using the --LogJoystick=1 command line option when starting up the Source Module. Thus, first, logging of joystick position needs to be enabled using that command line option. Then, the Expression Filter needs to be configured to simply replace the first and second control signals by the states that represent joystick position. This is accomplished by simply setting the parameter **Expressions** to the following 2-by-1 matrix:

```
JoystickXpos
JoystickYpos
```

By modifying the simple expression shown above, it would also be easy to use a combination of classified brain signals and states to drive an output such as the Cursor Task. You can find more information about the Expression Filter in Sect. 10.7.4.

References

1. Wilson, J.A., Mellinger, J., Schalk, G., Williams, J.: A procedure for measuring latencies in brain-computer interfaces. IEEE Trans. Biomed. Eng. (in press)

Part II
Technical Reference

Chapter 10
Core Modules

10.1 Operator

The operator module is the user interface seen by the experimenter. It makes it possible to view and edit system parameters, save and load parameter files, and to start and stop system operation.

10.1.1 Starting BCI2000

In addition to the operator module, a running BCI2000 system also consists of three core modules that must be started up together with the operator module. These core modules realize data acquisition, signal processing, and application. They can be started in two ways:

1. Using the batch scripts in the `BCI2000/batch` folder – choose one of the available scripts, or modify one to fit your needs. Script files provide an easy way to customize BCI2000 behavior using command line options, which are described in Sect. 6.3.1.
2. Using *BCI2000Launcher*. This tool provides a graphical user interface for module selection, and allows for automated loading of parameter files at startup.

10.1.2 Main Window

The main window shown in Fig. 10.1 contains four large buttons. These buttons correspond to the tasks that need to be performed during an experiment:

- *Config*: opens the configuration window,
- *Set Config*: applies the current set of parameters to the system,
- *Start*: starts system operation,

G. Schalk, J. Mellinger, *A Practical Guide to Brain–Computer Interfacing with BCI2000*,   135
© Springer-Verlag London Limited 2010

Fig. 10.1 The Operator module's main window

- *Quit*: exits the system once the experiment has finished.

Not all buttons are functional all the time; availability of a function depends on the state of the system. On top of the window, there are four function buttons (labeled Function 1–4) that execute Operator Scripts. Scripts and button captions may be configured freely from the preferences dialog. The status area, at the bottom of the main window, contains four fields. System status is displayed on the left, and module status is displayed in the three rightmost fields, in the natural order of core modules (i.e., Data Acquisition, Signal Processing, and Application).

10.1.2.1 Menus

The main window of the Operator module has three menus: *File, View,* and *Help*.

File Menu The *File* menu contains the following entries:

- *Preferences*: opens the Preferences dialog (see Fig. 10.2).
- *Exit*: terminates BCI2000 after displaying a confirmation message. This menu entry is always available, even if the *Quit* button is not enabled.

View Menu The *View* menu contains the following entries:

- *States*: displays a window that lists the state variables currently available in the system.
- *Operator Log*: toggles visibility of the log window. The log window is automatically opened when there are error or warning messages. You can display this window manually to view informative messages and debugging output.
- *Connection Info*: toggles visibility of the connection information window. This window displays information regarding whether connections between operator and core modules are established, network addresses, and the number of internal communication messages that have been transferred so far.

Help Menu The *Help* menu contains the following entry:

- *About*: displays extensive information about the operator module's version number, the source code revision it was built from, and the time of the build. This

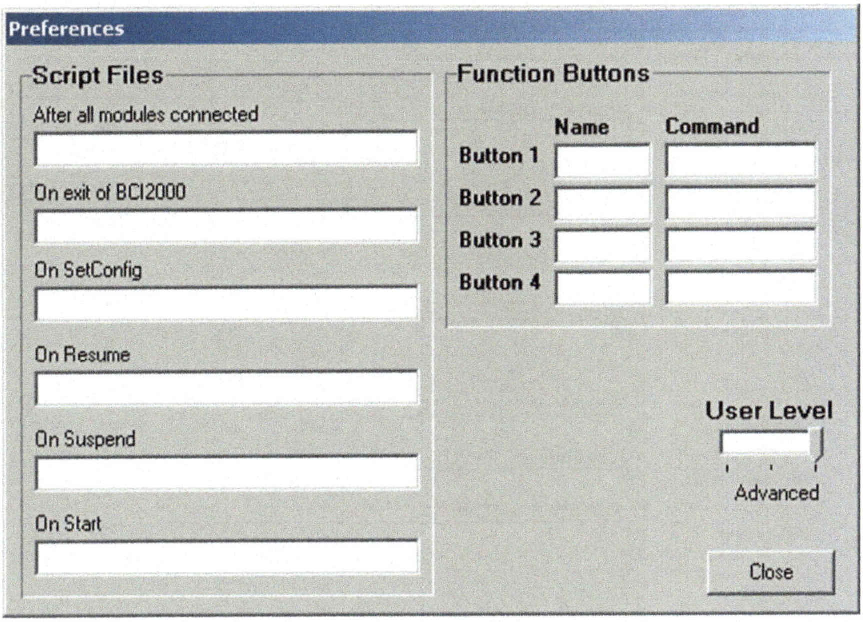

Fig. 10.2 The Preferences dialog

information is also kept inside the **OperatorVersion** parameter on the system tab, and written into data files along with all system parameters.

- *Help*: Opens a web browser window with the BCI2000 help system displayed.

10.1.2.2 The Preferences Dialog

On the left side, the preferences dialog allows you to associate operator scripts with events, and on the top right, you may configure the main window's function buttons using operator script commands. Please see Sect. 6.3.1 for scripting details. Scripts may also be specified from the command line at startup.

On the bottom right, you may specify the global **User Level** to be one of **beginner**, **intermediate**, or **advanced**. When the user level is set to **advanced**, a slider control is displayed for each individual parameter in the *Parameter Configuration Dialog* shown in Fig. 10.3. Each parameter is shown only when its user level is equal to or below the global setting. Thus, this feature allows to simplify the configuration dialog for less experienced users, or for your own convenience, in situations where full access to configuration options is not necessary. As a typical example, once you have configured a stimulation experiment, you may only need to change the **SubjectName** and **SubjectRun** parameters. Setting these two parameter's user levels to **beginner**, and leaving the other's at **advanced**, you may now switch the global user level to **beginner**, and will only see the two selected parameters displayed in the configuration dialog.

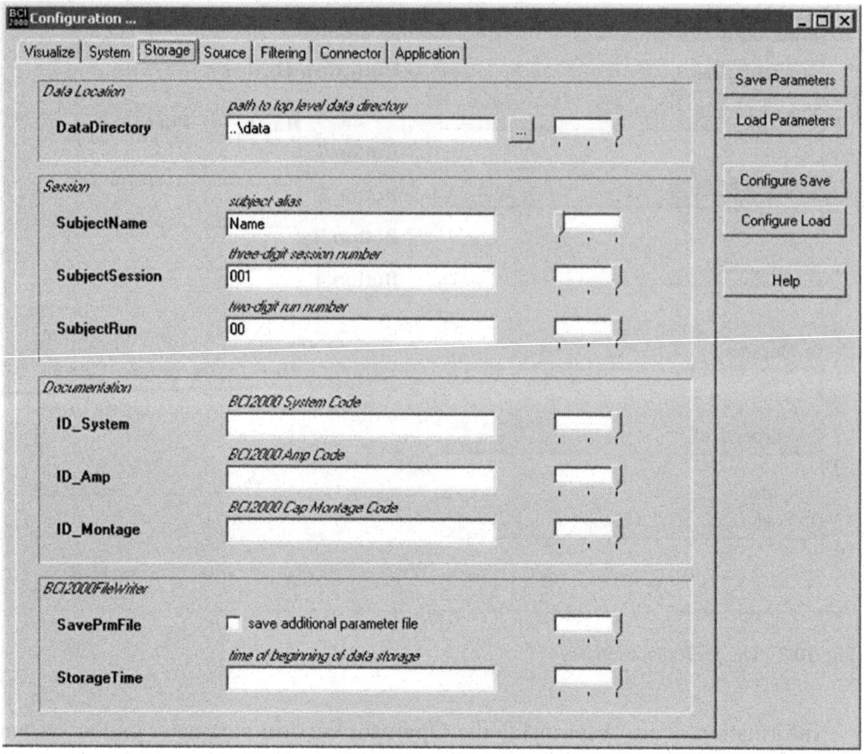

Fig. 10.3 The Parameter Configuration window

10.1.3 Parameter Configuration Window

The Parameter Configuration window is accessed by pressing the *Config* button in the main window. On the left side, this window displays a number of register tabs that correspond to parameter sections. Clicking on a tab will display the parameters from the corresponding section. Within a section, parameters are grouped according to subsections. In most cases, subsections correspond to individual filters (i.e., subcomponents of each module); however, filters with a large number of parameters might choose to group them into subsections to facilitate configuration. Within subsections, the order of parameter display depends on the order in which they are defined in the source code of the corresponding filter.

For parameters that hold one or more values, an edit field allows the user to change these values. For multiple values, entries are separated by a white space. If you would like to include values that contain white spaces, use "%20" (without the quotes) instead of those white spaces.

For some parameters, drop-down menus, checkboxes, or chooser buttons are displayed to make it more convenient for the user to enter parameter values. Matrix-valued parameters display a button to load or save a matrix in ASCII format, and an *Edit* button that opens up a *Matrix Editor* window.

On the right side, there are five buttons, *Load Parameters*, *Save Parameters*, *Configure Load*, *Configure Save*, and *Help*. The current settings in the **Configuration Window** can be saved to a BCI2000 parameter file (.prm) for future use with *Save Parameter*. The *Load Parameter* button opens a dialog to load either a parameter file (.prm) or data file (.dat). When you load a parameter file, the **Configuration Window** is filled with the parameter values of that parameter file. When you choose a data file, the **Configuration Window** is filled with the parameter settings that were used to record that data file. The latter possibility is useful to replicate the settings from a particular experiment.

It is important to note that the number and names of parameters in the parameter/data file do not exactly have to match those in the currently running BCI2000 implementation. As an example, the currently running BCI2000 implementation may be using a source module that realizes acquisition from device X (with some parameters that are specific to that device X), whereas the parameter file may have been created with a BCI2000 implementation that used a source module for device Y, and corresponding parameters whose names may be different than those for device X. In such a case, BCI2000 simply applies those parameters in the parameter/data file that are also present in the currently running BCI2000 implementation, ignores those that exist in the parameter file but not in the running BCI2000 implementation, and does not change those that exist in the currently running BCI2000 implementation but not in the parameter file. This feature is extremely useful when replicating settings from a different BCI2000 implementation (e.g., one with a different acquisition device). It also allows the use of so-called "parameter fragments." Parameter fragments contain only a subset of all parameters. For example, they may contain parameters that configure a data acquisition device for acquisition from a particular number of channels, or parameters that configure stimulus presentation for dual- or single-monitor configuration, etc. Because such fragments contain only a subset of all parameters, they are typically loaded on top of a full parameter file. You can create a parameter fragment by simply saving a parameter file, and then using a text editor to delete all parameters except those you would like to include.

Behavior of the load/save operation may be fine-tuned using the *Configure Load* and *Configure Save* buttons (Fig. 10.4). Each of these functions displays a list of parameter names and allows the user to select parameters that should be ignored during load/save operations. Thus, these functions essentially provide a different way to load/save parameter file fragments.

The parameter configuration window can also provide help to different parameters. When you click on the *Help* button, the mouse cursor changes into a question mark. Then, a click on a parameter name or edit field will open the appropriate help page when available. This help button is available only when a recent version of the BCI2000 doc directory is present along with the prog directory where the operator module resides.

Fig. 10.4 The Save (or
Load) Filter

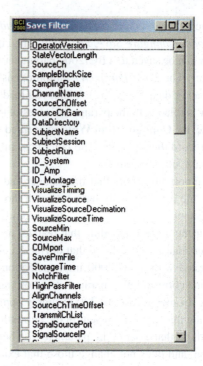

10.1.4 Matrix Editor Window

In BCI2000, certain parameters have values that are described as a matrix, i.e., they
consist of values that are organized into rows and columns. In addition, individual
matrix entries may be matrices themselves, allowing for multi-level nested configuration matrices. The matrix editor (see Fig. 10.5) allows you to:

- Modify the values in matrix-valued parameters by clicking an entry to edit it.
- Change the size of the matrix by changing the "number of rows" or "number of
  columns" fields, and clicking the "Set new matrix size" button.
- Edit row and column titles (labels) by right-clicking the matrix and choosing
  "Edit Labels" from the context menu to switch into label editing mode.
- Convert matrix entries into sub-matrices and back into single entries – use the
  context menu's "Convert to sub-matrix" and "Replace by single cell" items.
- Open another matrix editor to edit the contents of a sub-matrix – use the context
  menu's "Edit sub-matrix" item.

10.1.5 Visualization Windows

Visualization windows may be moved and resized; BCI2000 remembers their
positions across sessions. When you right-click on a visualization window (see
Fig. 10.6), a context menu will appear. It has several options that include:

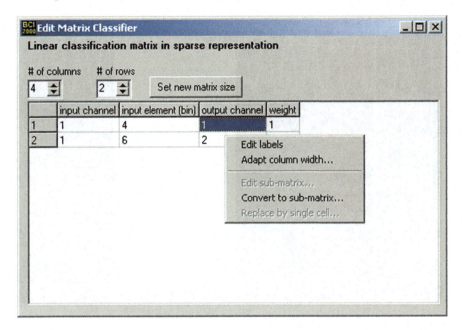

Fig. 10.5 The Matrix Editor window

- adjusting signal amplitude (enlarge/reduce signal),
- adjusting the time interval displayed in the window (fewer/more samples),
- choosing the number of signals displayed (fewer/more channels),
- switching between signal display as lines, and colored blocks ("display mode"),
- choosing signal colors,
- switching signal baselines, and unit display on and off,
- applying highpass, lowpass, or notch filters to the displayed signal.

For signal visualizations, the arrow keys and page up/page down keys may be used to scroll through channels. The full set of keyboard shortcuts is given in Table 10.1.

When the signal's physical unit is displayed, it is represented as a white bar (Fig. 10.7). In the example in this figure, there are white markers above and below the bar. These markers indicate the range corresponding to the value written in the bar. These markers may be missing; then the range corresponds to the height of the bar itself.

Finally, the physical unit may be displayed with a white marker attached to the bar (Fig. 10.8). This display is used for nonnegative signals. Here, the physical unit corresponds to the distance of the white marker to the signal's baseline.

Clicking *Choose Signal Colors...* from the context menu will open up a color dialog. This dialog has a field "Custom colors." The list of colors in this field defines the colors of the signal channels and is terminated with a black entry. Signal colors

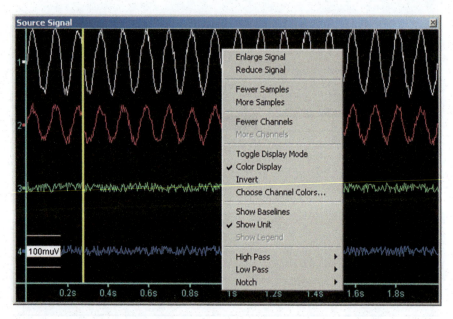

Fig. 10.6 A visualization window

Table 10.1 Keyboard shortcuts

up/down	Back/forward one channel
page-up/page-down	Back/forward one screenful of channels
-/+	Reduce/enlarge the signal
,/.	Fewer/more channels
left/right	Slow down/speed up the time sweep
home/end	Jump to the first/last screenful of channels
typing a number and pressing return (or 'g')	Jump to the specified channel number

are taken from these fields in the order in which they appear; when there are more signals displayed than colors defined, colors will be re-used in order.

10.2 Filter Chain

Each of the three BCI2000 core modules contains a *chain of filters*, i.e., a sequence of filters that form a filter chain. Each filter accepts data (e.g., raw brain signals) as input and produces an output (e.g., brain signal features).

Each filter produces an output each time it receives an input. That output is then automatically submitted to the subsequent filter for processing. This is analogous to a water pipe: unlike a water *stream*, it is impossible to insert or remove water from

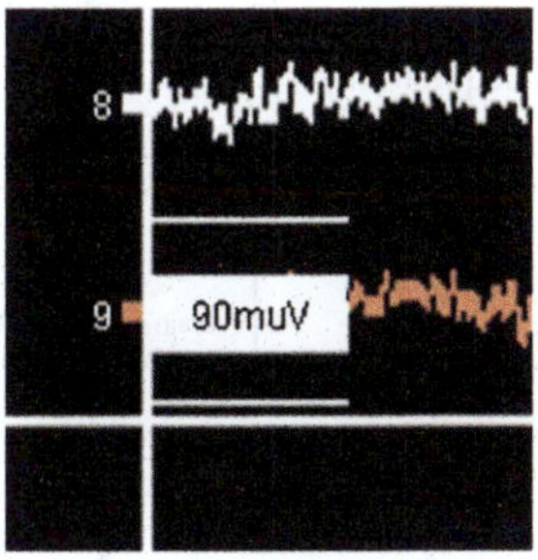

Fig. 10.7 The signal's physical unit is displayed inside a *white bar*, with *markers above and below* corresponding to the value written in the bar

Fig. 10.8 In this example, a *white marker* is attached to the unit's *bar*. The distance from 0 (i.e., in this case the x-axis) corresponds to the value written in the bar

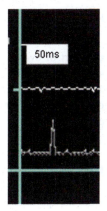

inside the *pipe* without breaking it. Similarly, although signal portions may change their shape on their way through the pipe (filter chain), it is impossible to insert or remove any of them. Thus, each data portion acquired by the data acquisition module will run through the entire BCI2000 system where it is processed by a sequence of filters.

10.3 Data Acquisition Filters

The source module contains several filters that handle data acquisition and storage, signal conditioning, and transmission to the signal processing module. These functions are realized by the **DataIOFilter**, **AlignmentFilter**, and **TransmissionFilter**, respectively (Fig. 10.9). The DataIOFilter realizes four functions: (1) data acquisition, which is implemented by an ADC filter; (2) data storage, which is realized by one of several FileWriter filters that support different data formats; (3) signal calibration into physical units (typically, µV); and (4) visualization of the acquired (brain) signals.

Fig. 10.9 Filters in the
source (i.e., data acquisition)
module

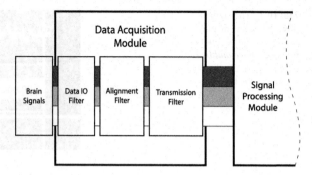

10.3.1 DataIOFilter

The **DataIOFilter** is part of every source module, and manages data acquisition, storage, and signal calibration into physical units (µV). For data acquisition, the DataIOFilter interfaces to an ADC component present inside a source module. For actual data storage, the DataIOFilter uses one of multiple **FileWriter** components present in a source module.

10.3.1.1 Parameters

Generally, data acquisition parameters are used for analog/digital (i.e., A/D) hardware configuration. Ideally, A/D hardware is configured by the ADC component; and thus, configuration changes to the hardware can simply be done by changing the parameter values in the Operator's config dialog. If the software interface to the A/D hardware does not allow for hardware configuration, actual A/D configuration must be done manually, and data acquisition parameters in the Source module have to be set to the same values. The following parameters are present in all BCI2000 Source modules, irrespective of the hardware device it supports:

ChannelNames A list of textual labels that are associated with channels. For EEG recordings, channel names should be electrode positions according to the 10–20 system (Fz, CPz, CP3, . . .). While not mandatory, providing labels is a means to document recording locations in the data file itself, so that this information does not need to be maintained externally (e.g., in a lab notebook). Also, BCI2000 makes it possible to refer to channels by their labels. Thus, channel labels make it easier to avoid errors in configuring feedback parameters.

DataDirectory Path to an existing directory. Recorded data will be stored below that directory. The path may be absolute, or relative to the source module's working directory at startup. Usually, the working directory at startup is the prog directory.

ID_System, ID_Amp, ID_Montage These parameters are provided for documentation purposes (i.e., of the system, amplifier, and montage, respectively) and may contain arbitrary text.

SampleBlockSize The number of sample points that make up a BCI2000 sample block. The temporal duration of such a sample block is given by the ratio of **SampleBlockSize** to **SamplingRate**, and determines timing resolution for feedback.

SamplingRate The number of samples per second per channel.

SourceCh The number of channels to be digitized and stored in the data file.

SourceChOffset, SourceChGain Calibration information defining conversion from AD units into physical units (μV). Raw data is converted into μV according to:

$$\text{calibrated}_{ch} = (\text{raw}_{ch} - \text{SourceChOffset}_{ch}) \times \text{SourceChGain}_{ch}$$

SourceMin, SourceMax For the source signal display, the minimum and maximum expected signal value in microvolts. SourceMin may exceed SourceMax to allow inversion of the signal display.

SubjectName A textual ID for the subject. The subject name will appear in session directories and data file names.

SubjectSession A textual ID for the current session. Sessions correspond to data directories. The path of a session directory is constructed according to:

$$\{\text{DataDirectory}\}/\{\text{SubjectName}\}\{\text{SubjectSession}\}$$

SubjectRun A number indicating the current run. Runs correspond to data files. To avoid accidental loss of data, run numbers are automatically incremented from the largest used value. Within a session directory, data file names are constructed according to:

$$\{\text{SubjectName}\}S\{\text{SubjectSession}\}R\{\text{SubjectRun}\}$$

A data file's extension depends on the output file format (see Sect. 6.3.3).

VisualizeTiming Switches display of timing information on or off.

VisualizeSource Switches display of the raw source signal on or off.

VisualizeSourceDecimation An integer decimation factor for the source signal display. For example, if you specify a factor of 2, only every other sample is drawn in the source display. This option is very important for configurations with high sampling rates, channel counts, and fast update rates, to ensure that BCI2000 can process incoming blocks fast enough.

VisualizeSourceTime The amount of time displayed in the source signal display.

10.3.1.2 States

SourceTime A 16-bit time stamp with a resolution of 1 ms, and wrap-around occurring every 65,536 ms. The time stamp is set immediately after a block of data has been acquired from the A/D converter.

StimulusTime A 16-bit time stamp in the same format as the **SourceTime** state. This time stamp is set immediately after the application module has updated the stimulus/feedback display.

Recording While data is being recorded this parameter is set to 1, 0 otherwise. In a data file, this state will always be 1.

Running 1 while data is being processed, 0 in suspended state. Setting this state to 0 from a filter, or over the **App Connector** interface (Sect. 6.4), will suspend BCI2000 operation.

10.3.2 AlignmentFilter

The AlignmentFilter performs temporal alignment of channel data. This is done using linear interpolation between subsequent time points.

Typically, the AlignmentFilter is used for A/D converters where channel sampling occurs subsequently rather than concurrently (e.g., in cases in which the device only has one sample-and-hold component). As a result, channels are time-shifted relative to each other by a considerable fraction of the sampling period. This time-shift must be corrected for to avoid adverse effects in certain filtering operations, such as spatial filtering.

Newer hardware, which is clocked at high speeds and employs oversampling to improve signal-to-noise ratio, may have no or only a negligible time-shift between channels.

10.3.2.1 Parameters

AlignChannels A nonzero value enables channel alignment.

SourceChTimeOffset A list of floating point values between 0 and 1. There must be one entry for each input channel. Each entry gives the amount of time-shift that is present for the respective channel, given in units of the sampling interval (i.e., [0, 1)). Alternatively, the list may be empty; then, an equidistant distribution of time shifts over the [0, 1) interval is assumed.

10.3.2.2 States

None.

10.3.2.3 Example

At a sampling rate of 250 Hz, the sampling interval is 1/(250 Hz) = 4 ms. If **SourceChTimeOffset** is an empty list, and if there are eight input channels, the following time shifts are assumed:

Channel	1	2	3	4	5	6	7	8
Relative time shift	0	1/8	1/4	3/8	1/2	5/8	3/4	7/8
Absolute time shift	0	0.5 ms	1 ms	1.5 ms	2 ms	2.5 ms	3 ms	3.5 ms

Entering the values given in the **relative time shift** row into **SourceChTimeOffset** would have the same effect as providing an empty list.

10.3.3 TransmissionFilter

The TransmissionFilter forwards a subset of input channels to its output. Typically, it is used in the data acquisition module to select a subset of recorded channels for on-line processing (i.e., to select those channels that are being submitted to the Signal Processing module). This option is used to reduce data transmission overhead and processing load.

10.3.3.1 Parameters

TransmitChList A list of input channels to be forwarded to the filter's output. Channels may be specified by ordinal number, or textual label if available. Channel labels are forwarded if present for the input signal.

10.3.3.2 States

None.

10.3.4 SourceFilter

Many biosignal amplifiers offer filtering in hardware. Generally, they have a line noise hum filter, or "notch filter," built in, and perform some high pass filtering. For amplifiers that do not provide these filters themselves, the *SourceFilter* realizes filtering in software.

Unlike other signal processing filters, the *SourceFilter* filter will be applied immediately after data acquisition, and it will modify the data which will be saved to disk, i.e., it will behave just as if it were a hardware filter built into the amplifier. All other signal processing filters are located in the Signal Processing module where they will be applied to the signal online but have no effect on the signal that is stored in the data file.

10.3.4.1 Parameters

HighPassFilter Configures the high pass filter, which is realized as a first order infinite impulse response (IIR) filter:

- 0: disabled,
- 1: at 0.1 Hz.

NotchFilter Configures the notch filter, which is realized as a 2×3rd order Chebychev bandstop. The proper setting depends on your country:

- 0: disabled,
- 1: at 50 Hz (Europe, Asia, Africa, parts of South America),
- 2: at 60 Hz (North America, parts of South America).

10.3.4.2 States

None.

10.3.4.3 Remarks

Generally, the *SourceFilter* will be available in source modules connecting to amplifiers that require it. Adding the *SourceFilter* to modules that do not yet contain it requires recompilation.

To add the *SourceFilter* to a new or existing source module, add the following files to the project, and recompile:

- `BCI2000/src/shared/modules/signalsource/SourceFilter.cpp`
- `BCI2000/src/shared/modules/signalprocessing/IIRFilterBase.cpp`
- `BCI2000/src/extlib/math/FilterDesign.cpp`

10.3.5 BCI2000FileWriter

The **BCI2000FileWriter** component stores data in the native BCI2000 file format.

10.3.5.1 Parameters

SavePrmFile Specifies whether parameters are to be stored in a parameter file separate from the data file. The additional parameter file will be redundant, as native BCI2000 data files always contain the complete set of parameters.

StorageTime At the beginning of the recording, this parameter's content is overwritten with a string describing the current date and time. Any user-specified value is ignored.

10.3.5.2 States

None.

10.3.6 EDF FileWriter

The **EDFFileWriter** component stores data in the EDF File Format.

10.3.6.1 Parameters

EquipmentID A string that identifies the equipment provider.
LabID A string that identifies the recording laboratory.
SignalUnit A string specifying the physical unit of the calibrated signal. Unlike the notation used throughout the rest of BCI2000, this is usually "uV" rather than "muV."
SubjectSex An enumerated value specifying the subject's sex:

- 0: not specified,
- 1: male,
- 2: female.

SubjectYearOfBirth Year the subject was born, in YYYY format.
TechnicianID A string that identifies the recording technician.
TransducerType A string that describes the transducer (sensor) type, e.g., "EEG: Ag/AgCl."

10.3.6.2 States

None.

10.3.7 GDF FileWriter

The **GDFFileWriter** component stores data in the GDF File Format.

In addition to the parameters provided by the **EDFFileWriter**, the GDF-FileWriter provides an option to define mappings between state variables and GDF events.

10.3.7.1 Parameters

SubjectYearOfBirth Year the subject was born, in YYYY format.
SubjectSex An enumerated value specifying the subject's sex:

- 0: not specified,
- 1: male,
- 2: female.

TransducerType A string that describes the transducer (sensor) type, e.g., "EEG: Ag/AgCl."

SignalUnit A string that specifies the physical unit of the calibrated signal. Unlike the notation used throughout the rest of BCI2000, this is usually "uV" rather than "muV."

EquipmentID A string that identifies the equipment provider. Since GDF version 2.0, the equipment ID field has been 8-byte numeric rather than string-valued. Consequently, the EquipmentID parameter is treated as an integer value, and written into the numeric field. If the parameter string evaluates to 0 – e.g., because it begins with a letter rather than a digit – its first 8 bytes are copied into the numeric equipment ID field in a byte-wise manner. Remaining bytes are padded with zero bytes if the parameter's string representation contains fewer than eight characters.

EventCodes BCI2000 does not prescribe the meaning of its state variables. GDF, on the other hand, associates a fixed set of events with certain numeric codes. Thus, a general mapping of BCI2000 states onto GDF events is not possible. Instead, GDF events are created via a user-defined set of mapping rules in the EventCode parameter, which also has a set of rules predefined for the most important cases.

The **EventCodes** parameter is a two-column matrix, with each row associating a boolean expression with a hexadecimal event code as defined in the GDF File Format. The expression is evaluated for each data block. Whenever its value switches from "false" to "true," the GDF event code from the second column is entered into the GDF event table; whenever its value switches back from "true" to "false," the same event code is entered into the table but with the most significant bit set to 1 (bitmask 0x8000).

The default value of the **EventCodes** parameter defines the most common mappings between BCI2000 states and GDF event codes:

Condition	Code	GDF event
TargetCode!=0	0x0300	trial begin
TargetCode==1	0x030c	cue up
TargetCode==2	0x0306	cue down
(ResultCode!=0)&&(TargetCode==ResultCode)	0x0381	hit
(ResultCode!=0)&&(TargetCode!=ResultCode)	0x0382	miss
Feedback!=0	0x030d	feedback onset

In addition to GDF events, BCI2000 state variables will be mapped to additional signal channels the same way as for EDF.

10.3.7.2 States

None.

10.3.8 NullFileWriter

The **NullFileWriter** component does not store brain signal data (null file format). Individual filters may still write log files to the directory defined by the **DataDirectory**, **SubjectName**, and **SubjectSession** parameters.

10.3.8.1 Parameters

SavePrmFile Specifies whether there should be a parameter file created for each run.

10.3.8.2 States

None.

10.4 Logging Input Filters

BCI2000 provides several mechanisms for recording from different user interfaces (e.g., mouse position, key presses, and joystick position and buttons) at sample resolution. The input logger is the preferred method of storing this data, but several older methods are maintained for compatibility. The Input Logger is described first, followed by the obsolete methods.

10.4.1 Input Logger

BCI2000 allows you to record input from keyboard, mouse, and joystick at sample resolution. This information is recorded into state variables, and recording is enabled by specifying appropriate command line options when starting a source module.

Input from keyboard, mouse, or joystick will occur asynchronously with respect to BCI2000's data processing. To account for this, input events are associated with time stamps, stored in a BCI2000 event queue, and assigned sample positions once data samples have been acquired.

When BCI2000 is distributed across multiple machines in a network, input devices must be attached to the machine that runs the data acquisition module. This is because time stamps of data acquisition and input events must refer to a common physical time base to allow associating input events with sample positions.

In Microsoft Windows, keyboard, mouse and joystick devices attached via USB are limited to a polling rate of 125 Hz, which corresponds to a temporal resolution of 8 ms. When better timing resolution is required, you might consider recording into an additional analog channel, or using the amplifier's trigger input when available.

The input logger is enabled using command line parameters.

10.4.1.1 Parameters

LogKeyboard When set to 1 from the command line, this parameter enables recording of keyboard events into the **KeyDown** and **KeyUp** state variables. Enabled with `--LogKeyboard=1`

LogMouse When set to 1 from the command line, this parameter enables recording of mouse events. Enabled with `--LogMouse=1`

LogJoystick When set to 1 from the command line, this parameter enables recording of joystick state. Enabled with `--LogJoystick=1`

10.4.1.2 States

KeyDown, KeyUp Key events. When a key is pressed, "KeyDown" will be set to the key's virtual key code at the corresponding sample position. When a key is released, the key code will be written into the "KeyUp" state variable.

MouseKeys Mouse key state, with the left mouse button corresponding to bit 0, and the right mouse button corresponding to bit 1.

MousePosX, MousePosY Mouse position in screen pixel coordinates with an additional offset of 32,768, i.e., the main monitor's top left corner will be recorded as (32,768, 32,768).

JoystickXpos, JoystickYpos, JoystickZpos Position is recorded from Joystick #1. Each position state ranges from 0 to 32,767 with a resting position at 16,384.

JoystickButtons1, JoystickButtons2, JoystickButtons3, JoystickButtons4 Joystick button information. Each button state is either 1 when pressed or 0 when not pressed.

10.4.2 JoystickFilter (Obsolete)

In contrast to input loggers, regular filters are only updated once per sample block. Thus, a filter that reads from an input device and stores the results in a state variable can detect changes in the input device only at a rate determined by the current sample block size. Thus, for detecting asynchronous external events, input loggers are preferred. In any case, several filters for input/output functionality exist. They include the **JoystickFilter**.

The **JoystickFilter** records joystick movements into a state value, making it possible to use it for data analysis purposes.

Note: The **JoystickFilter** is superseded by BCI2000's input logging facility, and kept for compatibility with existing experiments. Make sure not to activate input logging in a configuration where the **JoystickFilter** is present as well; otherwise, corruption of logged data may occur.

10.4.2.1 Parameters

JoystickEnable Enables recording of joystick movements when nonzero.

10.4.2.2 States

JoystickXpos, JoystickYpos, JoystickZpos Each position state ranges from 0 to 32,767 with a resting position at 16,384.
JoystickButtons1, JoystickButtons2, JoystickButtons3, JoystickButtons4
Each button state is either 1 when pressed or 0 when not pressed.

10.4.3 KeyLogFilter (Obsolete)

The **KeyLogFilter** logs keyboard events (i.e., key presses) and the state of mouse buttons. These events are recorded in BCI2000 states. This allows for tracking of user responses. Temporal resolution is limited to the length of a sample block. A keypress at any time during a sample block interval will result in that key's code being recorded for all respective state values within this sample block.

Note: The **KeyLogFilter** is superseded by BCI2000's input logging facility, and kept for compatibility with existing experiments. Make sure not to activate input logging in a configuration where the **KeyLogFilter** is present as well; otherwise, corruption of logged data may occur.

10.4.3.1 Parameters

LogKeyPresses Enables or disables keypress logging. Independent of the value of this parameter, the **MouseKeys** and **PressedKey** states will always be present in the file if the **KeyLogFilter** was added to the module.

10.4.3.2 States

MouseKeys A 2-bit state with bit 0 representing left mouse button state.

PressedKey1, PressedKey2, PressedKey3 These states contain "virtual key codes" (as defined by the Win32 API) for up to three keys. When more than three keys are pressed simultaneously, only those are recorded that were pressed first. The association of keys with states depends on the order in which keys were pressed. A state value of zero indicates that no key was pressed.

10.4.4 MouseFilter (Obsolete)

The **MouseFilter** captures the mouse position on the screen in device coordinates. Coordinates are always recorded.

Note: The **MouseFilter** is superseded by BCI2000's input logging facility, and kept for backward compatibility. Make sure not to activate input logging in a configuration where the MouseFilter is present as well; otherwise, corruption of logged data may occur.

10.4.4.1 Parameters

None.

10.4.4.2 States

MousePosX, MousePosY These states hold mouse cursor position in device coordinates, i.e., in units of screen pixels.

10.5 Signal Source Modules

10.5.1 SignalGeneratorADC

The *SignalGeneratorADC* filter generates a test signal suited for BCI purposes. The test signal is a sine wave overlayed with white noise and a DC component. Test signal amplitude may be controlled using the system's pointing device, which is useful in testing sensorimotor-based BCI setups. The signal's DC component may be modulated by the value of a state variable, which is useful in testing ERP type BCI setups.

10.5.1.1 Parameters

DCOffset An offset value common to all channels. May be positive or negative, and given in raw or calibrated units.

ModulateAmplitude Controls whether signal amplitude follows pointer (mouse, trackball, or similar) movement.

NoiseAmplitude White noise amplitude for all channels, given in raw or calibrated units. The noise itself is independent among channels.

OffsetMultiplier An arithmetic expression that evaluates to a multiplication factor applied to the DC offset as given by the **DCOffset** parameter. This makes it possible to create a test signal that depends on the value of state variables. For example, to create a simulated evoked response in the P300 Speller's copy spelling mode, enter **StimulusType** as the multiplier expression, and a nonzero value into the **DCOffset** parameter.

RandomSeed An initialization value for the system's random number generator. When a nonzero value is specified, the random number generator will always produce the same sequence, leading to identical pseudorandom noise, which is most useful for testing purposes. If this is zero, the random number generator will be initialized from system time, and thus lead to different noise each time.

SignalType Selects the output signal's numeric type. The following options are provided:

- 0: 16-bit signed integer (int16),
- 1: 32-bit IEEE floating point (float32),
- 2: 32-bit signed integer (int32).

SineAmplitude The test signal's maximum amplitude. Modulation will be linear in the range between zero and this maximum amplitude. A raw number refers to A/D units; when immediately followed with a voltage unit (such as μV, or mV), the number is interpreted according to the **SourceChOffset** and **SourceChGain** parameters.

SineChannelX Index of a channel that receives a test signal controlled by the pointer's (e.g., mouse's) horizontal position.

SineChannelY Index of a channel that receives a test signal controlled by the pointer's vertical position. If the index is zero, then the test signal is given on all channels.

SineChannelZ Index of a channel that receives a test signal controlled by mouse keys. For mouse key control, the pattern of pressed keys is translated into its corresponding 2-bit number. Signal amplitude is then proportional to this number.

SineFrequency Test signal frequency. When specified as a raw number, the frequency is given in terms of the sampling rate. Absolute frequencies may be given as decimal numbers or fractions when directly followed with Hz. Examples:

- 0.25 (half the Nyquist frequency),
- 10.5 Hz, or
- 20/3 Hz.

10.5.1.2 States

Any state may occur in the expression given in the **OffsetMultiplier** parameter.

10.5.2 gUSBampADC

The **gUSBampADC** filter acquires data from one or multiple **g.USBamp** am-
plifier/digitizer systems. **g.USBamp** is an amplifier/digitizer combination from
g.tec medical engineering GmbH/Guger Technologies OEG (http://www.gtec.at).
Support for this device in BCI2000 consists of two components: A BCI2000-
compatible Source Module (**gUSBampSource.exe**) and a command-line tool
(**USBampgetinfo.exe**).

10.5.2.1 Hardware

The g.USBamp consists of 16 independent 24-bit A/D converters that can sample at
up to 38.4 kHz per channel. Because there is one A/D converter for each channel,
one particular sample is digitized at the exact same time for each channel. This
is unlike with traditional A/D converter boards that only have one A/D converter.
BCI2000 has a feature that can align samples in time (parameter **AlignChannels**).
Because this feature is not needed in conjunction with the g.USBamp, it needs to be
turned off (i.e., **AlignChannels** needs to be 0).

10.5.2.2 Parameters

AcquisitionMode If set to **analog signal acquisition**, the g.USBamp records analog signal voltages (default). If set to **Calibration**, the signal output is a sine wave
   test signal generated by the g.USBamp (which can be used to verify correct func-
   tion/calibration). If set to **Impedance**, regular analog signal acquisition is preceded
   by an impedance test. This impedance test reports electrode impedances for each
   channel in kOhm. In impedance mode, the grounds and references are automatically connected in the amp internally (and the values of the parameters **Common-
   Ground** and **CommonReference** are ignored). In this mode, you need to connect
   the ground and reference to block D.
CommonGround This parameter determines whether the g.USBamp internally
   connects the GND inputs from all four blocks (A, B, C, and D) together. If enabled (default), then the signal ground only needs to be connected to one input
   block, e.g., block A. Otherwise, all GND inputs need to be externally connected.
CommonReference The same as **CommonGround**, except for the signal reference.

DeviceIDs List of serial numbers (e.g., UA-2007.01.01) of all devices. If you have more than one device, this list determines the order of the channels in the recording and in the data file. If only one device is connected, this parameter may be set to **auto**.

DeviceIDMaster Serial number (e.g., UA-2007.01.01) of the master device that is generating the clock for all other (i.e., slave) amplifiers. If you only have one device, this parameter has to equal **DeviceIDs** or may be set to **auto**. If you have more than one device, then this parameter represents the serial number of the device whose SYNC goes to the slaves, i.e., the only device that has a cable connected at SYNC OUT, but none connected to SYNC IN.

DigitalInput Turn on digital input. If turned on, the last sampled channel on each amplifier will contain sampled values of digital input 0 on the DIGITAL I/O input block on the back of the device. For example, if **SourceCh** is 8, then channels 1–7 will represent analog inputs, and channel 8 will represent the digital input. Thus, if **DigitalInput** is turned on, **SourceCh** and **SourceChDevices** may be a maximum of 17. Future versions of this module will support more than one digital input.

DigitalOutput Turn on digital output. If turned on, the digital output channel 0 is set low for the duration of a data block's acquisition, and set high at the end. This is primarily used for the BCI2000 Certification process, but may also be employed for other purposes, such as synchronizing acquisition with an external device.

DigitalOutputEx If this parameter contains an arithmetic or boolean expression, digital output 1 will be set high whenever this expression evaluates to true (nonzero) and low when the expression evaluates to false (zero). An example for such an expression is: `(( StimulusCode == 0 ) && Running ))`. In this example, the digital output is high whenever no stimulus is presented and the system is running.

FilterEnabled Choose 1 if you want a band-pass filter, and 0 if you don't. Because the g.USBamp is a DC amplifier, its signals may have a substantial DC offset that you may want to filter prior to storage and processing.

FilterHighPass High-pass frequency for band-pass filter. You need to query the amp for possible values. See description of the **USBampgetinfo** tool (page 205) for more info.

FilterLowPass Low-pass frequency for band-pass filter. See description of the **USBampgetinfo** tool (page 205) for more info. Please note that, because the g.USBamp has a 6.8 kHz hardware-based antialiasing filter, internally samples at a very high rate, and then downsamples the signal to the desired sampling rate, you will never experience aliasing even if you do not enable a low-pass filter.

FilterModelOrder Model order for band pass filter. See description of the **USBampgetinfo** tool (page 205) for more info.

FilterType Type of band pass filter:

- 1=CHEBYSHEV
- 2=BUTTERWORTH

NotchEnabled The notch filter suppresses power line interference at 50 or 60 Hz. Choose 1 if you want a notch filter, and 0 if you don't.

NotchHighPass Similar to **FilterHighPass**.

NotchLowPass Similar to **FilterLowPass**.

NotchModelOrder Similar to **FilterModelOrder**.

NotchType Similar to **FilterType**.

SamplingRate The sampling rate of all connected g.USBamps. If you would like to use a bandpass or a notch filter, there needs to be a filter configuration for that particular sampling rate (see the section on the **USBampgetinfo** tool, page 205). (g.tec Guger Technologies may be able to provide you with an updated driver configuration file if you need a different filter.)

The g.USBamp supports the following sampling rates: 32, 64, 128, 256, 512, 600, 1,200, 2,400, 4,800, 9,600, 19,200, and 38,400 Hz. All sampling rates are supported for one or more amplifiers. If you are sampling at high rates and from multiple amplifiers, the CPU may be overloaded depending on the speed of your computer and the BCI2000 configuration. In case you are experiencing problems (e.g., data loss, jerky display, etc.), increase the **SampleBlockSize** so that you are updating the system less frequently (usually, updating the system 20–30 times per second is sufficient for most applications), and increase **Visualize → Visualize-SourceDecimation**. This parameter will decrease the number of samples per second that are actually drawn in the *Source* display. For example, at 38,400 Hz, four amplifiers (64 channels), and a system update of 30 Hz, the computer would have to draw more than 73 million lines per second in the Source display! BCI2000 V3.0 also includes a new feature that automatically decimates the signal appropriately.

SignalType Defines the data type of the stored signal samples (int16 or float32). If the data type is int16, signal samples (which are produced by the amplifier in units of µV) are converted back into virtual A/D units (see **Additional Information** section below). If the data type is float32, the signals are stored in units of µV. In this case, **SourceChOffset** should be 0, and SourceChGain should be 1 (since the conversion factor from µV into µV is 1).

SourceChDevices The number of channels acquired from each device. If there is only one device, this parameter has to equal **SourceCh**. For example, ′16 8′ will acquire 16 channels from the first device listed under DeviceIDs, and 8 channels from the second device listed under DeviceIDs. Data acquisition always starts at channel 1. The sum of all channels (e.g., 24 in this example) has to equal the value of **SourceCh**.

SourceChList The list of channels that should be acquired from each device. The total number of channels listed should correspond to **SourceCh**. For more than one device, **SourceChDevices** determines how the **SourceChList** values are mapped to each device. For example if **SourceChDevices** = ′8 8′ and **SourceChList** = ′1 2 3 4 13 14 15 16 5 6 7 8 9 10 11 12′, then channels 1–4 and 13–16 will be acquired on the first device, and channels 5–12 will be acquired on the second device. These channels will be saved in the data file as 16 contiguous channels. The order of channels does not matter; i.e., ′1 2 3 4′ is the same as ′2 3 1 4′. The channels are always in ascending order on a single device. Channels may not be listed twice on a single device; e.g., entering ′1 2 3 4 5 6 7 1′ if **SourceChDevices** = ′8′ will result in an error. If this parameter is left blank (the default), then all channels are acquired on all devices.

10.5.2.3 States

None.

10.5.2.4 Additional Information

Unlike many other systems, the g.USBamp is a DC amplifier system that digitizes at 24 bit. Bandpass and notch filtering is performed on the digitized samples, resulting in floating point signal samples in units of μV. BCI2000 currently supports signed 16 bit integers, signed 32 bit integers and floating point numbers for its data storage. If **SignalType** is set to int16 or int32, the floating point values acquired from the g.USBamp have to be converted back into integers before they can be stored and transmitted to Signal Processing. This is done by the following transformation:

$$\text{sample}_{\text{stored}}(\text{A/D units}) = \frac{\text{sample}_{\text{acquired}}(\mu V)}{\text{SourceChGain}} \qquad (10.1)$$

In other words, when using the g.USBamp and storing data in 16/32 bit integers, it is important to select values for **SourceChGain** such that the resulting integers are in appropriate value ranges. This is particularly important for 16 bit integers that only have a limited resolution (0–65,535). Otherwise, signal clipping or loss of resolution will occur. In summary, unless storage or other considerations require the use of integer signals, we strongly recommend to store signals from the g.USBamp in floating point representation.

BCI2000 Signal Processing or any offline analysis routine can derive, as with any other BCI2000 source module, sample values in μV by subtracting, from each stored sample, **SourceChOffset** (i.e., zero), and multiplying it with **SourceChGain** for each channel. If **SignalType** is set to float32, data samples are stored in units of μV. In this case, **SourceChGain** should be a list of 1's (because the conversion factor between data samples into μV is 1.0 for each channel).

When values other than 0 and 1 are specified, a consistent data file will be produced, i.e., values will be transformed before they are written to the file, such that applying SourceChOffset and SourceChGain will reproduce the original values in μV.

10.5.3 gMOBIlabADC

The gMOBIlabADC acquires data from a **g.MOBIlab device**. The **g.MOBIlab** is an amplifier/digitizer combination from g.tec medical engineering GmbH/Guger Technologies OEG (www.gtec.at).

10.5.3.1 Hardware

The g.MOBIlab device supports eight analog input channels digitized at 16 bit resolution and sampled at a fixed 256 Hz sampling rate. In its standard configuration, channels 1–2 have a sensitivity of ±100 µV, channels 3–4 have a sensitivity of ±500 µV, channels 5–6 have a sensitivity of ±5 mV, and channels 7–8 have a sensitivity of ±5 V.

The input range of the A/D converter is approximately equal to this sensitivity. Thus, for example, one LSB for channel 1 or 2 is roughly $\frac{200 \text{ µV}}{65,536} = 0.003$ µV. However, the actual input range of the A/D converter is slightly larger than the sensitivity of each channel (so that the A/D converter can detect when the amp saturates), and thus, exact LSB values have to be determined for each channel using a calibration signal.

This device has only one A/D converter. Thus, it digitizes signals from different channels at slightly different times. BCI2000 has a feature that can align samples in time (parameter **AlignChannels**, which needs to be enabled (i.e., **AlignChannels** needs to be 1)).

Additional features of the g.MOBIlab are two digital input/output lines. The g.MOBIlab source module is configured such that channel 9 corresponds to the value of the digital lines, which are configured as input lines.

10.5.3.2 Parameters

COMport Serial port of the attached g.MOBIlab device, e.g., COM2 :
SamplingRate The sampling rate of the g.MOBIlab. This value has to be 256.
SourceCh The total number of channels. This number can be 1 to 9. If it is set to 9, then channels 1–8 represent eight analog input channels, and channel 9 represents the values of the two digital lines.

10.5.3.3 States

None.

10.5.4 *gMOBIlabPlusADC*

The gMOBIlabPlusADC acquires data from a **g.MOBIlab+** device. The **g.MOBIlab+** is an amplifier/digitizer combination from g.tec medical engineering GmbH/Guger Technologies OEG (http://www.gtec.at) which can transmit data over a serial connection or wirelessly via a Bluetooth connection.

10.5.4.1 Hardware

The g.MOBIlab+ device supports eight analog input channels digitized at 16 bit resolution and sampled at a fixed 256 Hz sampling rate. Additionally, eight digital input lines can be sampled with the analog data, so that behavioral information (e.g., button presses) can be recorded as well. The amplifier has a sensitivity of ± 500 μV. Thus, one LSB is roughly $\frac{1\,\text{mV}}{65,536} = 0.0153$ μV.

This device only has one A/D converter. Thus, it samples signals from different channels at slightly different times. BCI2000 has a feature that can align samples in time (parameter **AlignChannels**, which needs to be enabled (i.e., **AlignChannels** needs to be 1)).

Additional features of the g.MOBIlab+ are four digital input lines, and four digital input/output lines. The g.MOBIlab+ source module is configured such that channels 9–16 correspond to the value of the digital lines, which are configured as input lines. It is possible to configure one digital line in the output configuration, so that the line is pulsed during a data read. This is described further in the parameters section below.

If using Bluetooth, g.tec Bluetooth software must be installed prior to using the g.MOBIlab+. Please follow the instructions provided by g.tec.

10.5.4.2 Parameters

COMport Serial port of the attached g.MOBIlab+ device, e.g., COM7: This value is determined when the Bluetooth serial port is configured during installation.

SourceCh The total number of channels. This number can be 1 to 8, or 16. If it is set between 1 and 8, then channels 1–8 represent eight analog input channels. If it is set to 16, then channels 1–8 are analog, and channels 9–16 are digital inputs (see **DigitalEnable**).

SampleBlockSize Samples per digitized block. A value of 8 corresponds to a BCI2000 system update rate of 32 Hz.

SamplingRate The sampling rate of the g.MOBIlab+. This value has to be 256.

InfoMode Displays information about the g.MOBIlab+ device.

DigitalEnable If this is set to 1, then the eight digital lines are read as inputs. In this case, the total number of channels (SourceCh) must equal 16.

DigitalOutBlock If this is set to 1, then digital line 8 is set to output mode, and is set low at the onset of block acquisition, and set high after the block is read. This allows the system timing to be measured, or to synchronize BCI2000 with an external device.

10.5.4.3 States

None.

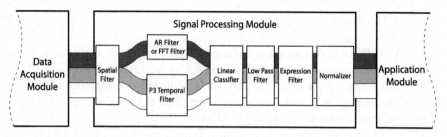

Fig. 10.10 Filters in the signal processing module

10.6 Signal Processing Filters

In the signal processing module, brain signals are processed by a series of filters that translate raw brain signals into device control signals. Thus, these filters realize feature extraction and translation. Feature extraction consists of filtering the signal spatially and temporally, which results in a set of extracted features. Signal translation consists of a linear classifier, which translates these features into control signals that differentiate amongst a small number of mental states (classes). Signal translation also includes the normalizer, which adjusts the classifier's output to zero mean and unit variance.

The sequence of these filtering procedures is shown in Fig. 10.10. The filters themselves are described in more detail below.

10.6.1 SpatialFilter

The **SpatialFilter** computes an instantaneous linear transformation of its input. Typically, the **SpatialFilter**'s input is the unfiltered brain signal from the source module. The linear transformation that is applied by the spatial filter is described by a transformation matrix, and applied for each sample separately, i.e., not linking data across different points in time. This linear transformation can be parameterized in three different ways as described below.

Each spatial filter type uses a different algorithm to compute the linear transformation, and can therefore have implications on CPU load and performance. In the following, N denotes the filter's number of input channels, and M denotes its number of output channels. Typically, M is less than or equal to N.

10.6.1.1 Parameters

SpatialFilterType This parameter defines the method that will be used to implement the spatial filter. Choices are:

0: None No spatial filtering is performed; the input signal is copied to the output signal, and the spatial filter matrix is ignored.

1: Full Matrix The linear transformation applied to the input signal is defined by the SpatialFilter matrix parameter. This is also the default, and matches behavior in previous versions of BCI2000.

2: Sparse Matrix The sparse matrix filter uses the SpatialFilter matrix parameter to define non-zero matrix entries. Each non-zero entry is given by an input channel, output channel, and weight for that channel.

3: Common Average Reference (CAR) The common average reference spatial filter calculates the mean of all channels, and subtracts this value from the selected output channels. This filter's output channels may be defined by the **SpatialFilterCAROutput** parameter.

SpatialFilterCAROutput This parameter is a list of channels that define which channels should be output from the common average reference spatial filter, and the order in which they should appear. That is, the location of the channel in this list determines the output channel position. For example, if input channels 6, 7, 10, 12 should be passed to the output of the spatial filter as channels 3, 4, 1, 2, then this parameter should be set to: [10 12 6 7]. Rather than numbers, channel names may be specified as defined in the **ChannelNames** parameter: e.g., [C3 C4 CP3 CP4 Cz].

10.6.1.2 States

None.

10.6.1.3 Spatial Filter Remarks

Full Matrix

The full matrix filter uses the **SpatialFilter** parameter to define the linear transformation applied to the filter's input signal. In this matrix, columns represent input channels, and rows represent output channels. Each matrix element defines a weight with which the respective input channel (column) enters into the respective output channel (row).

If the spatial filter should realize an identity filter – not modifying its input – then the **SpatialFilter** matrix should be set to a unit matrix (square matrix with ones on the main diagonal, and zeros in all other elements).

In a typical EEG experiment with a fixed montage, you might want column labels to reflect the respective electrode location, simplifying the task of further modifications to the spatial filter. Also, specifying row labels to identify output channels allows you to use those labels in configuration of further stages of processing, such as the **LinearClassifier**.

Full matrix representation is the most general way to specify a spatial filter, and recommended for situations not allowed for by the remaining spatial filter types. Compared to those, it is associated with a number of disadvantages:

- Complexity is $O(NM)$. In situations with large sampling rates and/or a large number of channels, this may result in real-time problems due to high CPU load.
- Channel labels are not preserved across the filter. Due to the filter's general nature, no relation between input and output channel names can be inferred, and all output labels must be specified manually.
- The filter's specification depends on the order of input channels, which makes it interdependent with the **TransmitChList** parameter.

Sparse Matrix Filter Type

The sparse matrix filter uses the **SpatialFilter** parameter to define the relationship between input channels and output channels with a given weight. In this case, the **SpatialFilter** matrix must have three columns, and a row for each input/output relationship. The first column contains the input channel, and the third column defines the weight that the input channel is multiplied by before being assigned to the output channel, which is defined in the second column.

When specifying input channels in the first column, you may use channel names. In order to assign labels to output channels in sparse representation, use arbitrary labels in the (second) output channel column.

The sparse matrix method performance is determined completely by the number of elements (rows) in the **SpatialFilter** matrix. In the best case, a single channel is multiplied by the weight and assigned to the specified output channel; this would take far less CPU time than the CAR method, and possibly the "none" option as well.

In a realistic scenario, all N input channels will be used to calculate $M \leq N$ output channels; thus, complexity is between $O(N)$ and $O(NM)$, with $O(N^2)$ in the worst case.

Sparse matrix worst-case performance will be close to that of an $N \times N$ "full" spatial filter matrix. Thus, for spatial filter configurations that depend on electrode/sensor locations such as Laplacian filtering, filter definition in terms of a sparse matrix using channel labels appears most advantageous.

Common-Average Reference

It is important to note that all channels passed to the spatial filter (typically defined in the **TransmitChList** parameter) are used in the CAR calculation, but only a subset of these channels are actually output and passed to the next step in the signal processing chain. If this parameter is left blank, then all input channels are passed to the output, and the number of input channels equals the number of output channels.

The common-average reference has a complexity of $O(N + M)$, providing the next best performance in most circumstances. It is possible to create a CAR using either the full-matrix or sparse matrix options; however, the CAR method only calculates the mean value once per sample, and subtracts it only from the selected output channels. In order to implement a CAR in a full-matrix, the mean must be recalculated for every output channel, which is not as efficient, particularly for high channel count systems.

10.6.2 ARFilter

The ARFilter computes an autoregressive model of its input (i.e., spatially fil-tered brain signals in an unmodified BCI2000 system) using the Maximum Entropy Method (Burg algorithm). Spectral estimation is done separately for each of the fil-ter's input channels. Its output may be raw AR coefficients, or an estimated power or amplitude spectrum collected into bins. Thus, it can be used in place of the **FFT-Filter**, which also computes a power spectrum.

10.6.2.1 Parameters

For all frequency-valued parameters, float values without any unit are interpreted as fractions of the sampling rate; absolute frequencies may be given when followed with Hz, as in the following examples:

- 0.5 refers to the Nyquist frequency (half the sampling rate).
- 10.3 Hz specifies a value of 10.3 Hz, regardless of the sampling rate.
- 15/2 Hz specifies a value of 7.5 Hz.
- The unit must immediately follow the number.

BinWidth A single non-negative float value that represents the width of a single bin, e.g., "3 Hz."

Detrend Determines whether signal samples are detrended prior to spectral estima-tion. Possible values are

- 0: no detrending,
- 1: mean removal,
- 2: linear trend removal.

EvaluationsPerBin A single non-negative integer value that represents the number of uniformly spaced evaluation points that enter into a single bin's value.

FirstBinCenter A float value representing the center of the first frequency bin, e.g., "5 Hz."

LastBinCenter A float value representing the center of the last frequency bin.

ModelOrder The order of the autoregressive model. This value approximately cor-responds to the maximum number of peaks in the resulting spectrum.

OutputType Possible values are:

- 0: spectral amplitude,
- 1: spectral power,
- 2: AR coefficients.

If the output is a spectrum, the output signal's elements correspond to uniformly spaced frequency bins. For raw AR coefficients, the first output signal element contains total signal power, followed by the model's denominator coefficients.

WindowLength The length of the input data window over which the model/spectrum is computed, given as a time value in seconds (then immediately fol-lowed by the unit), or the number of signal blocks (e.g., 1.34s, 500ms, or 5).

10.6.2.2 States

None.

10.6.2.3 ARFilter Remarks

AR coefficients are actually the coefficients of an all-pole linear filter that is fitted to reproduce the signal's spectrum when applied to white noise. Thus, the estimated power spectrum directly corresponds to that filter's transfer function, divided by the signal's total power. To obtain spectral power for finite-sized frequency bins, that power spectrum needs to be multiplied by total signal power, and integrated over the frequency ranges corresponding to individual bins. This is done by numerical integration, evaluating the spectrum at evenly spaced evaluation points, summing, and multiplying by bin width to obtain the power corresponding to a certain bin. For amplitude rather than power spectrum output, bin integrals are replaced with their square roots.

10.6.2.4 Example

For a typical EEG application, you might use the following configuration:

```
ModelOrder= 16
FirstBinCenter= 0Hz
LastBinCenter= 30Hz
BinWidth= 3Hz
EvaluationsPerBin= 15
```

This will result in 11 bins, with the first bin covering the DC range from -1.5 to 1.5 Hz. Due to symmetry of the transfer function, this is twice the integral from 0 to 1.5 Hz. The last bin will cover the range from 28.5 Hz to 31.5 Hz. Evaluation points within each bin are 0.2 Hz apart.

10.6.3 FFTFilter

The **FFTFilter** applies a short-term Fast-Fourier Transformation (FFT) to selected channels of its input signal, resulting in a time series of frequency spectra. The computed spectra may be displayed in visualization windows. Typically, the **FFTFilter** is used for spectral estimation and demodulation, instead of the **ARFilter**.

10.6.3.1 Parameters

FFTInputChannels A list of input channels for which the FFT is computed. When **FFTOutputSignal** is set to a value other than 0, **FFTInputChannels** list entries determine the correspondence between input and output channels.

FFTOutputSignal Depending on configuration, the **FFTFilter**'s output signal will be the computed spectrum, or the unchanged input. Possible values are

0: input connect-through This option allows using the **FFTFilter** for visualization purposes.

1: power spectrum As with the **ARFilter**, the output signal's elements will correspond to frequency bins.

2: complex amplitudes The output will be complex Fourier coefficients in half-complex format, with the spectrum's imaginary part appended to the real part.

FFTWindowLength The length of the input data window over which the FFT is computed, given as a time value in seconds (then immediately followed by the unit), or the number of signal blocks (e.g., 1.34s, 500ms, or 5). The FFT will be computed once per data block. If the length of the input data window exceeds that of a data block, FFT windows will overlap. If the data window is shorter than a data block, only the most recent samples will enter into the FFT.

FFTWindow Selects the type of side-lobe suppression window. Possible values are

1 Hamming window,
2 Hann window,
3 Blackman window.

VisualizeFFT A nonzero value selects visualization of the FFT-computed power spectrum. Independently of the **FFTOutputSignal** parameter's value, it is always the power spectrum that is visualized.

10.6.3.2 States

None.

10.6.3.3 FFTFilter Remarks

The **FFTFilter** requires the FFTW3 library which, for licensing reasons, you need to obtain separately from BCI2000. The **FFTFilter** uses the FFTW3 library to do the actual FFT computation, which is released under the GNU public license (GPL). Please download the windows version of FFTW 3.01 from http://www.fftw.org/install/windows.html.

In order to make the FFTW library available to BCI2000, put FFTW3.DLL into your BCI2000 installation's prog directory. When using the command line version of the **FFTFilter**, put a copy of FFTW3.DLL into your BCI2000 installation's tools/cmdline directory.

For recent Windows builds of FFTW3, the naming scheme has changed. If you would like to use such a build with BCI2000, put a copy of the file `libfftw3-3.dll` into `BCI2000/prog` and rename it to `FFTW3.DLL`. This is currently untested.

10.6.4 P3TemporalFilter

The **P3TemporalFilter** is often used instead of the ARFilter or FFTFilter. It performs buffering and averaging of epochs of data as required for online classification of evoked responses (ERPs). It averages over epochs of data that are recorded in response to a number of stimuli; averaging is done separately for each stimulus. When a fixed number of epochs have been accumulated for a given stimulus, the **P3TemporalFilter** computes their temporal average, and reports the average wave form(s) in its output signal. Whenever a temporal average is reported, the **StimulusCodeRes** state is set to the stimulus code that the average is in response to.

Typically, the **P3TemporalFilter**'s output is sent into the **LinearClassifier** which takes waveform data from multiple locations and time points to linearly combine them into a single output. This number can represent the size of the evoked response for each stimulus. Application modules such as the **StimulusPresentation** and **P3Speller** can then provide feedback on this output.

10.6.4.1 Parameters

EpochLength Determines the duration of an epoch. An epoch begins with the onset of its associated stimulus. Epoch length may be specified in blocks as a raw number, or in seconds when immediately followed with a unit: e.g., 500ms.

EpochsToAverage Determines the number of epochs that will be accumulated before reporting their average.

TargetERPChannel For waveform visualization, selects the channel to present in the average waveform display. Channels may be given by ordinal number, or textual label.

VisualizeP3TemporalFiltering If nonzero, average waveforms will be presented graphically to the investigator.

10.6.4.2 States

StimulusCodeRes When reporting a waveform average, this state contains the associated stimulus code. A **StimulusCodeRes** value of 0 indicates that the **P3TemporalFilter**'s output does not contain a valid waveform average, and should be ignored by the application module.

StimulusTypeRes When reporting a waveform average, this state is 1 if the associated stimulus was marked as "attended" on presentation, and 0 if the stimulus was not marked as "attended." In other words, the value of **StimulusTypeRes** at the time of waveform reporting matches that of **StimulusType** at the time of stimulus presentation.

10.6.4.3 States Used for Input

StimulusBegin (optional) When this state is present, it is assumed that a nonzero value indicates stimulus onset. When there is no **StimulusBegin** state present, stimulus onset will be derived from the **StimulusCode** state.

StimulusCode The code of a stimulus being presented. When there is no **StimulusBegin** state present, a switch of **StimulusCode** from zero to a nonzero value indicates stimulus onset. For averaging, epochs are grouped according to this state's value at stimulus onset.

StimulusType This state indicates whether a stimulus is marked as *attended*. Its value is stored along with a stimulus presentation's associated wave forms.

10.6.5 LinearClassifier

The ARFilter, FFTFilter, or P3TemporalFilter (or any other filter you may have written) extract signal features from spatially filtered brain signals. The **LinearClassifier** subsequently translates these features into output control signals using a linear equation. Thus, each control signal is a linear combination of signal features. Input data has two indices (N channels $\times$ M elements), and output data has a single index (C channels $\times$ 1 element); thus, the linear classifier acts as a $N \times M \times C$ matrix, determining the output after summation over channels and elements:

$$\text{output}_k = \sum_{i=1}^{N} \sum_{j=1}^{M} \text{input}_{ij}\, \text{Classifier}_{ijk} \tag{10.2}$$

In a BCI based on periodic brain signal components (such as a mu rhythm-based BCI), the **LinearClassifier**'s input is the time-varying amplitude or power spectrum for a number of spatially filtered EEG channels. Its output is normalized with respect to mean and variance, and then used as a control signal to determine cursor movement.

In an ERP-based BCI (such as the P300 BCI), the **LinearClassifier**'s input is a sequence of averaged EEG time courses obtained in response to a number of stimuli, and its output is considered to represent the likelihood for each of these responses to be the desired evoked response.

10.6.5.1 Parameters

Classifier The **Classifier** parameter is a sparse matrix definition in which each row corresponds to a single matrix entry. Columns correspond to

1. input channel,
2. input element (bin in the spectral case, time offset in the ERP case),
3. output channel,
4. weight (value of the matrix entry).

Input channels may be specified by ordinal number, or by textual label if available (e.g., CP4). Input elements may be given as ordinal number, or using appropriate units (e.g., 10Hz, or 120ms).

10.6.5.2 States

None.

10.6.5.3 Mu Rhythm Classification Examples

For these examples, let us assume that you set the **ARFilter**'s **FirstBinCenter** to 0, and **BinWidth** to 3Hz. This allows you to refer to the respective bins by their frequency, i.e., 12Hz rather than 5. We also assume that, in the **SpatialFilter**, you have entered labels for output channels. This allows you to refer to channels by their names, i.e., CP3 rather than 7.

1D Cursor Movement
In this example, you want to give feedback from electrode CP4, using the amplitude between 10.5Hz and 13.5Hz for cursor feedback. Then, the **Classifier** parameter will have a single row that contains

Input channel	Input element	Output channel	Weight
CP4	12 Hz	1	1

2D Cursor Movement
In this example, you want to use mean activity from the left and right hand areas at 12Hz to control movement in horizontal direction. Additionally, you want the difference between left and right hand areas at 24Hz to control movement in vertical direction. In the **CursorTask** application, the horizontal direction (X) corresponds to channel 1, and the vertical direction corresponds to channel 2. Accordingly, your **Classifier** parameter will have the following four rows:

Input channel	Input element	Output channel	Weight
CP3	12 Hz	1	0.5
CP4	12 Hz	1	0.5
CP3	24 Hz	2	−0.5
CP4	24 Hz	2	0.5

Note that, in the above example, subtracting CP3 from CP4 in the classifier is not equivalent to taking the difference in the **SpatialFilter**. This is because (at least for a mu rhythm-based BCI) spatial filtering is followed by computation of spectral amplitudes, and feature combination amounts to an addition of spectral amplitudes. Computing spectral amplitudes involves taking the absolute value (or its square), which is a nonlinear operation, where typically $|A - B| \neq |A| - |B|$.

10.6.5.4 ERP Classification Example

Often, you will use a computer program to optimize the classifier for use with ERPs (such as P300 GUI or P300 Classifier). However, for the sake of an instructive example, let us assume that you want to classify based on the temporal mean between 280 and 300ms and the spatial mean from electrodes Cz and Pz, and that your sampling rate is 250Hz. Then, there will be six samples in that range, starting with sample number 70:

Input channel	Input element	Output channel	Weight
Cz	70	1	1
Cz	71	1	1
Cz	72	1	1
Cz	73	1	1
Cz	74	1	1
Cz	75	1	1
Pz	70	1	1
Pz	71	1	1
Pz	72	1	1
Pz	73	1	1
Pz	74	1	1
Pz	75	1	1

Note that, unlike in the case of spectral features, there is no difference between spatially combining channels in the **LinearClassifier** or in the **SpatialFilter**. Thus, you might as well combine Cz and Pz into a channel labeled Cz+Pz in the **SpatialFilter**, and then use this classifier configuration:

Input channel	Input element	Output channel	Weight
Cz+Pz	70	1	1
Cz+Pz	71	1	1
Cz+Pz	72	1	1
Cz+Pz	73	1	1
Cz+Pz	74	1	1
Cz+Pz	75	1	1

10.6.6 Normalizer

The Normalizer applies a linear transformation to its input signal, i.e., the control signals produced by the classifier, so that the output control signal is in a certain value range. For each channel (index denoted with i), the normalizer subtracts an offset and multiplies the result with a gain value. As described later, these offset and gain values can either be manually defined or can be determined automatically during online BCI2000 operation using an adaptation procedure:

$$\text{output}_i = (\text{input}_i - \text{NormalizerOffset}_i) \times \text{NormalizerGain}_i \qquad (10.3)$$

If enabled, the **Normalizer** estimates offset and gain values adaptively from the past statistics of its input in an attempt to make its output signal zero mean and unit variance. The **Normalizer** uses "data buffers" to accumulate its past input according to user-defined rules. These rules are called "buffer conditions," because they are given in terms of boolean expressions. Each data buffer is associated with such a boolean expression. Whenever an expression evaluates to "true," the current input will be appended to the associated buffer. Whenever it comes to updating offset and gain values, the **Normalizer** will use the content of its buffers to estimate data mean and variance. The offset will then be set to the data mean, and the gain to the inverse square root of the data variance, i.e., the inverse of the data standard deviation.

10.6.6.1 Parameters

For each channel of the **Normalizer**'s input signal, adaptation is treated independently. Offsets, gains, and choice of adaptation are represented as list parameters, where each entry in the list corresponds to a signal channel. Buffer configuration is done in matrix form, where columns correspond to signal channels and rows correspond to adaptation buffers.

Adaptation A list of values that determines adaptation strategy individually for each input channel (i.e., control signal). Possible values are

- 0 for no adaptation,
- 1 for adjusting offsets to zero mean,
- 2 for adjusting offsets to zero mean and gains to unit variance.

BufferConditions A matrix that consists of boolean expressions. Expressions may involve state variables and the components of the **Normalizer**'s input signal (see examples below). Each matrix entry represents a data buffer that is a ring buffer of length **BufferLength**. Whenever a buffer's expression evaluates to true, the current value of the input signal will be put into the buffer (overwriting its oldest data once the buffer is filled). Columns correspond to control signal channels. Buffers in a certain column will buffer data from the corresponding signal channel, and will be used in adaptation of that channel only. Within columns, the order of buffers does not affect computation. Empty expressions do not have any effect on the computation. Thus, it is possible to have a different number of buffers for different channels. As an example, a buffer to store data for the first target in the Cursor Task and during feedback only, should have an expression such as `(Feedback)&&(TargetCode==1)`.

BufferLength The maximum length of each data buffer. The length is specified in data blocks if given as a raw number, and in seconds if given as a number followed by the character `s`. All data buffers have the same capacity. Once a data buffer is filled, its older entries will be replaced with new data (ring buffer). In previous versions of BCI2000, buffer lengths were specified in terms of "past trials." This enforced the notion of a "trial," and did not generalize to continuous adaptation.

NormalizerOffsets, NormalizerGains Lists of offset and gain values where each entry corresponds to one control signal channel. These values will be updated depending on the channel's adaptation configuration in the **Adaptation** parameter.

Update Trigger A boolean expression that triggers adaptation when changing to `true` from `false`. Generally, continuous adaptation within trials is not desired. Usually, adaptation will be set to occur at the end of a trial. This is achieved with **UpdateTrigger** expressions such as `Feedback==0` or `TargetCode==0`. For continuous adaptation, specify an empty string in the **UpdateTrigger** parameter.

10.6.6.2 States

Buffer condition expressions and the **UpdateTrigger** expression may involve any states present in the system. Expressions are checked for syntactical correctness and whether the states that are referred to in these expressions are also present in the system.

Adaptation Rationale
It may appear crude to use the total data variance for the adaptation – why not use linear regression on data labels (target codes) to separate user controlled (task determined) from noise variance? User controlled variance would then correspond to target separation on the feedback screen, which is what we want to normalize in the first place.

However, a closer look reveals that the "relative size" of user controlled variance, and noise variance is crucial. When that "signal" variance is small compared to noise variance, we would be ill advised to use it in normalization – this would only lead to enlarged noise, and an erratically moving cursor on the feedback screen. In this case, we rather want to normalize by noise variance, to keep the cursor well-behaved. At the same time, total variance approaches noise variance in this limit because signal variance is small.

On the other end of the spectrum, we have a signal variance that is large compared to noise variance. Here, we clearly want normalization by signal variance. However, the total variance will be dominated by signal variance. Thus, in the limit of high signal-to-noise ratio, total variance again is the quantity by which we want to normalize. Thus, no matter whether signal-to-noise ratio is high or low, total data variance appears to be a good choice for normalization.

Finally, it would also be possible to adaptively estimate the signal dynamics of the signal features (such as spectral amplitudes), rather than simply adapting the classifier output, i.e., control signal values. While this seems appealing mathematically and has been shown to produce valuable results [7, 8], it also requires estimation of a potentially large number of features. This implies that the resulting system will generally be less stable, that adaptation will take much longer, and that appropriate function of such an adaptation scheme will require careful configuration by an expert. Consequently, we chose the procedure described above for signal adaptation within BCI2000.

Typical Use
The **Normalizer**'s input is the output of the **Classifier**. The **Normalizer** transforms this input into a control signal with zero mean and unit variance. This control signal is then transmitted to an application module. Because the application module can assume that each control signal has zero mean and unit variance, it can easily relate these control signals to task-specific parameters such as window size, screen update rate, cursor velocity, and trial duration.

10.6.6.3 Examples

Trial-Based 1D Feedback Task with Three Targets
In this example, only data from the feedback phase should enter into the adaptation. To make sure that targets contribute equally to the adaptation, we use a single buffer for each target. We use a 3-rows-by-1-column **BufferConditions** matrix (see below). Depending on which control signal should be adapted on, all of these conditions need to be entered in the first, second, or third column of the matrix (corresponding to x, y, and z direction when used with the CursorTask application module, respectively).

```
(Feedback)&&(TargetCode==1)
(Feedback)&&(TargetCode==2)
(Feedback)&&(TargetCode==3)
```

We want to use data from approximately three previous trials of each target. The feedback duration (i.e., the duration of cursor movement) is 2 seconds. Thus, we set the buffer length to the equivalent of three feedback durations:

```
BufferLength= 6s
```

Adaptation should happen at the end of each trial, when feedback is finished. Thus, we set **UpdateTrigger** to an expression that changes to "true" when feedback ends:

```
UpdateTrigger= (Feedback==0)
```

To achieve continuous movement in, say, the x direction, we need a constant normalizer output on that channel. To be consistent with the task module's **Feedback-Duration** parameter, this output should be constant at $+2$ in order for the cursor to cross the entire screen during a trial, and $+1$ when cursor movement begins at the screen's center. This corresponds to the following settings for the channel in question: **Adaptation** $= 0$, **NormalizerOffset** $= -1$, **NormalizerGain** $= 1$ or 2, respectively.

Trial-Based 2D Feedback Task with Four Targets
Leaving everything else as in the previous example, we now have two dimensions corresponding to left-right (channel 1) and up-down (channel 2). The target positions are as indicated below:

```
- - - - - - - - - - - - - - - - - - - - - - -
|             ##1##              |
|#                              #|
|2                              3|
|#                              #|
|             ##4##              |
- - - - - - - - - - - - - - - - - - - - - - -
```

We use data from targets 1 and 4 to adjust channel 2, and targets 2 and 3 to adjust channel 1. Accordingly, the **BufferConditions** matrix is

```
(Feedback)&&(TargetCode==2)  (Feedback)&&(TargetCode==1)
(Feedback)&&(TargetCode==3)  (Feedback)&&(TargetCode==4)
```

Continuous 1D Control Without Pre-defined Targets
For this example, we assume that, over a period of 10 minutes, all output values will occur with approximately equal frequency, or at least have a symmetric distribution around zero. The difference from the previous examples is that we here do not assume any knowledge about position and timing of targets. This example could realize operation of BCI2000 with the Dasher spelling system, or with other devices that expect statistically balanced input. The **BufferConditions** matrix will have a single entry containing a constant expression:

```
1
```

This way, data will always be buffered. There are no trials. We want a continuous adaptation to the values of the last 10 minutes. Thus, we set the **BufferLength** parameter to $600\,s$, or $(10*60)\,s$. For continuous adaptation, we enter an "empty string" (not a constant 0 expression) for **UpdateTrigger**.

10.7 Additional Signal Processing Filters

10.7.1 LPFilter

The LPFilter is a simple single-pole temporal low-pass filter with a time constant T (given in units of a sample's duration), a sequence $S_{in,t}$ as input and a sequence $S_{out,t}$ as output (where t is a sample index proportional to time), where

$$S_{out,0} = \left(1 - e^{-1/T}\right) S_{in,0}$$

$$S_{out,t} = e^{-1/T} S_{out,t-1} + \left(1 - e^{-1/T}\right) S_{in,t}$$

Typically, the **LPFilter** is used to remove high-frequency noise from the **Classifier**'s output, i.e., to smooth the resulting control signals.

10.7.1.1 Parameters

LPTimeConstant The filter's time constant in units of sample blocks, or in seconds if followed with an s. Consistent with its mathematical limit, a value of **LPTime-Constant**= 0 disables the **LPFilter**.

10.7.1.2 States

None.

10.7.2 ConditionalIntegrator

The **ConditionalIntegrator** filter accumulates (integrates) its input signal over time, depending on the value of a given boolean expression. Typically, the **ConditionalIntegrator** is used in off-line analysis of data collected in a cursor movement paradigm. During on-line operation, the feedback cursor's position represents the integral of the control signal. For off-line parameter simulation experiments, the **ConditionalIntegrator** performs this integration, resulting in an output equivalent to cursor position on the feedback screen. A typical feedback task sets the Feedback state to 1 during cursor movement. Accordingly, one would use IntegrationCondition= Feedback in an off-line simulation.

10.7.2.1 Parameters

IntegrationCondition A boolean expression that determines whether the signal is integrated. When the expression first evaluates to true, the filter's output is set to zero. Then, while the expression evaluates to true, filter input is integrated, and the integral's current value is returned as the filter output. When the expression first evaluates to false, filter output will be kept at the last integration value, and will stay there until the expression again evaluates to `true`.

10.7.2.2 States

Any existing state may be part of the **IntegrationCondition** expression.

10.7.3 StateTransform

The **StateTransform** filter replaces state values with values of an arithmetic expression. Typically, the **StateTransform** filter is used as a functional equivalent for cursor hit detection in off-line data analysis, using the output of the **ConditionalIntegrator** filter to determine cursor position. When the cursor hits a target, a typical cursor movement task sets the **ResultCode** state to the number of the target that was hit. For the case of 1D feedback with two targets, and when the filter's input is the integrated control signal as output by the **ConditionalIntegrator**, the following **StateTransforms** matrix achieves this behavior:

State	Expression
ResultCode	$(ResultCode > 0) * (1 + (Signal(1, 1) > 0))$

10.7.3.1 Parameters

StateTransforms Any number of transforms, specified in the form of an $(N \times 2)$ matrix. Each row corresponds to a transform. Within a row, the first column gives the name of the state to be transformed, and the second column gives the expression that replaces that state's value. The expression may also involve the value of the state it replaces.

10.7.3.2 States

Any state may be replaced, and any state may occur in the replacement expression.

10.7.4 ExpressionFilter

The **ExpressionFilter** uses arithmetic expressions to determine its output signal. As arithmetic expressions may contain state variables and the filter's input signal, this provides a powerful way to modify data processing according to BCI2000 system state, to introduce state variable information into additional signal channels, or to replace processing results with system state information.

10.7.4.1 Parameters

Expressions A matrix of strings that represent arithmetic expressions. These expressions determine the filter's output signal. Output signal dimensions are derived from matrix dimensions:

- matrix rows correspond to signal channels, and
- matrix columns correspond to signal elements (samples).

For each signal block, each of the expressions is evaluated, and the result is entered into the corresponding element in the output signal. Expressions may contain references to the filter's input signal, and to state variables (see examples below). When the **Expressions** parameter is an empty matrix, the **ExpressionFilter** will simply copy its input into its output.

10.7.4.2 States

Any available state may enter into one of the expressions.

10.7.4.3 Examples

The following 2-by-2 matrix will replace a 2-by-2 input signal with its square. Additionally, inserting `*(ResultCode==0)` into each entry will change the output to zero when the **ResultCode** state variable is nonzero.

```
Signal(1,1)^2 Signal(1,2)^2
Signal(2,1)^2 Signal(2,2)^2
```

When the **Input Logger** is used to track a joystick position, the **JoystickXpos** and **JoystickYpos** state variables represent joystick position. When the **ExpressionFilter** is present in the signal processing module, placed between **LinearClassifier** and **Normalizer**, it may be configured to control the cursor directly by joystick position, using this 2-by-1 matrix:

```
JoystickXpos
JoystickYpos
```

In the same configuration, to re-enable cursor movement by the output of signal processing without removing the **ExpressionFilter** from the signal processing module, either set the **Expressions** parameter to an empty matrix, or use this matrix:

```
Signal(1,1)
Signal(2,1)
```

10.7.5 MatlabFilter

The Matlab filter allows you to implement the algorithms of your choice in Matlab. Once BCI2000 is running with the Matlab filter, you will see that a Matlab command line window has opened. In that command line window you can type commands that show you the variables that BCI2000 communicates to the Matlab engine. For example, you may enter one of the following commands:

```
% show the variables
whos

% plot the first channel of the data (see below)
plot(bci_InSignal(1,:))

% plot the first channel of the data and continuously
% update the plot (see below)
while(1); plot(bci_InSignal(1,:)); pause(0.01); end
```

If you try these examples above like this, you will notice that Matlab takes quite a considerable amount of time to open a new figure. While the figure is being opened by Matlab, the engine is blocked, which causes the MatlabFilter to fail writing new data to the engine. Hence the MatlabFilter will return an error and you will have to restart BCI2000. The reason is that the Matlab engine gives you access through the command window, but the Matlab engine is not capable of executing the commands that it receives from BCI2000 and the commands that it receives from you through the command window simultaneously.

To get the above example to run correctly, you will have to open the figure prior to starting BCI2000 by typing figure (this will open an empty figure). Subsequently, once BCI2000 is running, the content of the figure can be updated.

However, the above example already shows that the Matlab engine is not intended to be used concurrently with the BCI2000 MatlabFilter and from the command window. Instead of typing your own commands in the command window, you will have to give full control of command execution to the MatlabFilter to ensure that timing is controlled. The way that BCI2000 and Matlab interact is that each of the components of a standard filter is mapped onto a corresponding Matlab function. If the Matlab filter is running within BCI2000, it will call the Matlab engine and execute the required Matlab function. The easiest way of getting started with the Matlab filter is by using the tutorials on IIR bandpass filtering followed with RMS envelope computation and linear classification in Sect. 7.4.

10.7.5.1 Parameters

The Matlab filter does not have a set of default parameters that it uses. Instead, the user-supplied Matlab functions that are executed by the Matlab filter specify the parameters. After initialization, these parameters are displayed in the Operator and can be modified there.

10.7.5.2 States

None.

10.7.5.3 Troubleshooting

Matlab doesn't find your functions Make sure to either set Matlab's working directory to the directory that contains your functions, using the `--MatlabWD= <path>` command line option, or add the respective directory to your Matlab path. We generally recommend the first option.

There is no Matlab engine started up Execute `matlab/regserver` from the command line when logged in with administrative privileges.

10.8 Application Modules

10.8.1 Cursor Task

The purpose of the Cursor Task is to provide a BCI2000 user application module that can realize 1D, 2D, or 3D cursor movement tasks. It displays a box-shaped scene in which a ball-shaped cursor moves and is controlled by the output of the *Signal Processing Module*. In this scene, targets appear as cuboids or rectangles. The box, cursor, and targets may be texturized.

The Cursor Task implements a cursor movement task based on a three-dimensional control signal communicated to it by a BCI2000 Signal Processing module. This task progresses through five subsequent stages. These stages are illustrated in the timeline shown in Fig. 10.11. In stage 1, the screen shows an empty

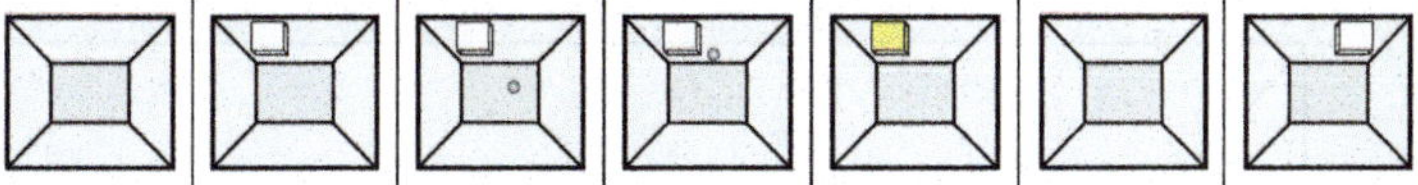

State							
TargetCode	0	2	2	2	2	0	3
Feedback	0	0	1	1	0	0	0
IntertrialInterval	1	0	0	0	0	1	0
ResultCode	0	0	0	0	1	0	0
Stage	1	2	3	4	5	6	7

Fig. 10.11 Timeline for the Cursor Task

workspace. This period is called Inter-Trial Interval. Subsequently, a target (that is represented as a cuboid) appears in one out of n possible locations. This period is called Pre-Trial Pause. After the pre-trial pause, the cursor appears in stage 3. It immediately starts to move as determined by the 3-dimensional control signal. During stages 3 and 4, the user's task is to move the cursor towards and into the target. This period of cursor movement is called the Feedback Period or Trial Period. Period 4 can end in one of three ways: Either the cursor hits the correct target, it misses by hitting any other of the defined targets, or the feedback period takes too long and times out. Stage 5, the Reward Period, follows stage 4. The cursor disappears and the target changes its color to indicate completion of the trial period. After a defined duration, the target disappears and the next trial starts with an Inter-Trial Interval.

Visual Representation

The visual representation of the Cursor Task consists of a three-dimensional workspace (see Fig. 10.12) that can (if selected) be indicated by five bounding rectangles. These rectangles can have a user-selectable texture, which is the same for each rectangle. Targets are represented by cuboids, whose edges can also have a particular texture. Finally, the cursor is represented by a sphere. It, too, may have a user-defined texture. To facilitate depth perception, the cursor's color provides an additional cue about the cursor's Z position as follows. The operator specifies the cursor's color at the top-most and at the bottom-most Z positions. For any position between these extremes, the cursor's color will be linearly interpolated between the two colors. (Specifically, each of the three color components (i.e., red, green, and blue) will be interpolated to result in the cursor's color for a given Z position.)

Control Signal

Cursor movement is controlled by the output of the Signal Processing Module. In this signal, channels 1, 2, and 3 correspond to dimensions X, Y, and Z, and there is a single value (element) present in each channel, defining cursor velocity in the respective dimension. Additional channels or elements that might be present in the control signal will be ignored by the CursorTask application module.

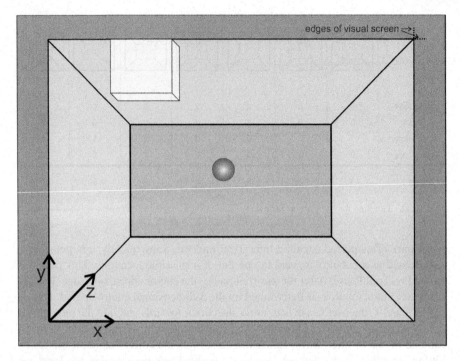

Fig. 10.12 Visual display of the Cursor Task

10.8.1.1 Parameters

Positions and sizes are given in a percent coordinate system. There, [0 0 0] corresponds to the bottom left corner of the workspace, and [100 100 100] corresponds to its diagonal counterpart.

CameraAim Camera aim point in percent coordinates.
CameraPos Camera position vector in percent coordinates of the 3D area.
CameraProjection An enumerated value that specifies one of the following camera projections:

- 0: flat,
- 1: wide angle perspective,
- 2: narrow angle perspective.

CursorColorFront, CursorColorBack The cursor's color when it is in the front or back of the workspace, given as an RGB value. When different colors are specified for front and back, the cursor's z coordinate will be used to linearly interpolate between the two color values.
CursorPos The cursor's starting position, given as a vector in percent coordinates.
CursorTexture The path to a texture file to be used for the cursor.
CursorWidth The feedback cursor's width, given in percent of screen width.

FeedbackDuration Typical duration of feedback. Given in sample blocks, or in time units when immediately followed with 's,' 'ms,' or similar. This parameter is not a hard limit to feedback duration but determines cursor velocity such that, for a normalized control signal, cursor movement will take the specified time from the cursor's starting point to the screen's edge. Feedback trials will typically have this duration, provided that the cursor starts at the center of the screen, targets are located at the screen's edges, and cursor and target width are ignored (for further details, see the section on the Normalizer above).

ITIDuration The duration of the inter-trial interval. Given in sample blocks, or in time units when immediately followed with 's,' 'ms,' or similar.

LightSourcePos Light source position in percent coordinates.

LightSourceColor The light source's color in RGB encoding.

MaxFeedbackDuration Abort a trial after this amount of feedback time has expired. Unlike the FeedbackDuration parameter, this is a hard limit. Given in sample blocks, or in time units when immediately followed with 's,' 'ms,' or similar.

MinRunLength The duration of a run, i.e., the time corresponding to a continuously recorded data file. A run will not stop during a trial, so its actual length may be larger than this value by the length of a trial. If this parameter is set, then **NumberOfTrials** should be blank, and vice-versa. Given in sample blocks, or in time units when immediately followed with 's,' 'ms,' or similar.

NumberTargets The number of targets on the feedback screen.

NumberOfTrials The number of trials in a run. If this parameter is set, then **MinRunLength** should be blank, and vice-versa.

PreFeedbackDuration The duration of target display before feedback begins. Given in sample blocks, or in time units when immediately followed with 's,' 'ms,' or similar.

PreRunDuration The duration of the pause preceding the first trial. Given in sample blocks, or in time units when immediately followed with 's,' 'ms,' or similar.

PostFeedbackDuration The duration of result display after feedback. Given in sample blocks, or in time units when immediately followed with 's,' 'ms,' or similar.

RenderingQuality An enumerated value with one of the following entries:

- 0: low – 2D rendering. Lighting, shading, and textures are switched off, even if specified otherwise for individual objects. Also, collision detection ignores objects' z positions. On machines without OpenGL compatible 3D hardware, 2D rendering is considerably faster than 3D rendering.
- 1: high – 3D rendering. Lighting, shading, and textures are applied as specified.

TargetColor Target color in RGB encoding.

Targets A matrix with six columns. The first three columns define the position coordinates of the target's center, given in percent coordinates; the last three columns define the target's three-dimensional extent (i.e., width, height, depth, respectively). Each row corresponds to one target. Targets are always cuboids aligned with the three coordinate axes.

TargetTexture Path to a texture file to be used for targets, or empty for none. Currently, Windows BMP files are accepted as textures. Paths may be absolute, or relative to the executable's working directory at startup, which usually matches the executable's location.

TestAllTargets An enumerated value that determines collision test behavior:

- 0 to test only the visible current target,
- 1 to test all targets.

WindowBitDepth The feedback window's color bit depth.

WindowWidth, WindowHeight The width and height of the subject-visible application window, in pixels.

WindowLeft, WindowTop The screen position of the application window's top left corner, in pixels.

WorkspaceBoundaryColor The workspace boundary color in RGB encoding. The special value of 0xff000000 hides the workspace boundary.

WorkspaceBoundaryTexture The path to a workspace boundary texture, or empty. Currently, Windows BMP files are accepted as textures. Paths may be absolute, or relative to the executable's working directory at startup, which usually matches the executable's location.

10.8.1.2 States

CursorPosX, CursorPosY, CursorPosZ These states record the cursor's position, translated into the 0–4,095 range such that the 3D scene's left, top, and down planes all correspond to 0, and the right, bottom, and up planes correspond to 4,095.

Feedback This state's value is 1 when the cursor is displayed on the feedback screen. Typically, this also implies that the cursor moves according to the control signal.

ResultCode At the end of a feedback trial, **ResultCode** is set to the target code of the outcome, i.e., the target that was hit by the cursor. When a time of **PostTrial-Duration** has passed, **ResultCode** is reset to zero.

StimulusTime A 16-bit time stamp in the same format as the **SourceTime** state. This time stamp is set immediately after the application module has updated the stimulus/feedback display.

TargetCode During a feedback trial, this state indicates the target that is visible and that the cursor is supposed to hit. A **TargetCode** value of zero indicates that no target is on the screen. When **TargetCode** switches from zero to nonzero, the trial begins.

10.8.2 Stimulus Presentation

The main purpose of this task is to present a sequential series of auditory and/or visual stimuli to the user. Thus, this **StimulusPresentationTask** is suited to implement a wide array of studies including evoked response (ERP) paradigms or motor movement/imagery sessions. In addition to stimulus delivery, the task can also be used in conjunction with BCI2000's P300 Signal Processing module (P3SignalProcessing.exe) to provide feedback to a selected stimulus in either a copy or a free mode.

10.8.2.1 Parameters

AudioSwitch A boolean parameter to globally switch presentation of audio stimuli on or off. To present audio for individual stimuli only, remove audio entries for other stimuli from the **Stimuli** matrix.

AudioVolume The volume for audio playback, in percent of maximum volume. This parameter's value may be overridden by an additional row in the **Stimuli**, **FocusOn**, and **Result** matrices.

BackgroundColor The background color of the stimulus rectangle, given as an RGB value. The height of the rectangle is defined by the **CaptionHeight** parameter, and its width depends on the caption's text width.

CaptionColor For text stimuli, the color of the stimulus' caption, given as an RGB value. This parameter's value may be overridden by an additional row in the **Stimuli**, **FocusOn**, and **Result** matrices, as described further below.

CaptionHeight For text stimuli, the height of the stimulus' caption in percent of screen height. This parameter's value may be overridden by an additional row in the **Stimuli**, **FocusOn**, and **Result** matrices.

CaptionSwitch A boolean parameter to globally switch presentation of stimulus captions on or off. To present captions for individual stimuli only, remove captions for other stimuli from the **Stimuli** matrix.

DisplayResults Switches result display of copy/free spelling on or off.

FocusOn In copy mode (see **InterpretMode**), the attended stimulus is presented prior to the stimulus sequence (**PreSequenceTime**), preceded with a special announcement stimulus. This stimulus' properties are defined by the **FocusOn** parameter, which is a matrix in the same format as the **Stimuli** parameter. Usually, this matrix has a single column; when multiple columns are present, all stimuli are presented concurrently.

IconSwitch A boolean parameter to globally switch presentation of icon stimuli on or off. To present icons for individual stimuli only, leave icon entries for other stimuli blank in the **Stimuli** matrix.

InterpretMode An enumerated value that determines whether this task should be used in conjunction with the P3SignalProcessing module for on-line classification of evoked responses:

- 0: no target is announced "attended," signal processing classification results are ignored;
- 1: online or free mode: classification is interpreted to show the selected stimulus, but no "attended target" is defined;
- 2: copy mode: "attended" targets are defined, classification is interpreted, and selected stimulus is shown.

ISIMinDuration, ISIMaxDuration Minimum and maximum duration of the inter-stimulus interval. During the inter-stimulus interval, the screen is blank, and audio is muted. Actual inter-stimulus intervals vary randomly between minimum and maximum value, with uniform probability for all intermediate values. Given in sample blocks, or in time units when immediately followed with 's,' 'ms,' or similar. Note that temporal resolution is limited to a single sample block.

NumberOfSequences The number of sequence repetitions in a run (a run corresponds to a single data file).

PostRunDuration Duration of the pause following last sequence. Given in sample blocks, or in time units when immediately followed with 's,' 'ms' or similar.

PostSequenceDuration Duration of the pause following sequences (or sets of intensifications). Given in sample blocks, or in time units when immediately followed with 's,' 'ms,' or similar. When used in conjunction with the **P3Temporal-Filter**, this value needs to be larger than the **EpochLength** parameter. This allows classification to complete before the next sequence of stimuli is presented.

PreRunDuration The duration of the pause preceding the first sequence. Given in sample blocks, or in time units when immediately followed with 's,' 'ms,' or similar.

PreSequenceDuration Duration of the pause preceding sequences (or sets of intensifications). Given in sample blocks, or in time units when immediately followed with 's,' 'ms,' or similar. In free or copy mode, the **PreSequenceDuration** and **PostSequenceDuration** parameters may not go below twice the value of the **StimulusDuration** parameters, in order to allow for presentation of **FocusOn** and **Result** announcement stimuli.

Result In copy and free modes (see **InterpretMode**), the classification result is presented following the sequence (**PostSequenceTime**). Presentation of the predicted stimulus is preceded with an announcement stimulus. This stimulus' properties are defined by the **Result** parameter, which is a matrix in the same format as the **Stimuli** and **FocusOn** parameters.

Sequence In deterministic mode, a list of stimulus codes that define the sequence of presentations. In random mode, a list of integer stimulus frequencies.

SequenceType Enumerated value that can be:

- 0: deterministic sequence mode: the sequence is explicitly defined in the Sequence parameter;
- 1: random sequence mode: the sequence is random, with pre-defined stimulus frequencies.

Stimuli A matrix that defines the stimuli and their properties. Columns of the Stimuli matrix correspond to individual stimuli and their stimulus codes. For each stimulus, the following properties are defined by its row entries:

- Caption: a text string, with its size and color depending on the **CaptionHeight** and **CaptionColor** parameters;
- Icon: a graphic file (Windows BMP), with its size depending on the **StimulusWidth** parameter;

- Audio: an audio file (Windows WAV), with playback starting at the onset of the visual stimuli.

A blank entry for caption/icon/audio is accepted, and implies that no presentation of the respective element takes place.

Additionally, a number of global stimulus parameters may be overridden with specific values for individual stimuli. To do this, for each parameter to be individualized, add an additional row to the **Stimuli** matrix. The row label indicates the parameter to be changed, and has to be one of:

- **StimulusDuration,**
- **ISIMinDuration, ISIMaxDuration,**
- **StimulusWidth, CaptionHeight, CaptionColor, AudioVolume.**

Whenever one of these rows is present, the corresponding global parameter will be ignored.

StimulusDuration For visual stimuli, the duration of stimulus presentation. For auditory stimuli, the maximum duration, i.e., playback of audio extending above the specified duration will be muted. Given in sample blocks, or in time units when immediately followed with 's,' 'ms,' or similar.

StimulusWidth For icon stimuli, stimulus width in percent of screen width. **Stimulus** height is deduced from the stimulus' aspect ratio, which is always conserved. If this parameter is zero, all stimuli will be displayed unscaled, i.e., at their original pixel size. This parameter's value may be overridden by an additional row in the **Stimuli**, **FocusOn**, and **Result** matrices.

ToBeCopied A list of stimulus codes that define a sequence of attended stimuli. At the beginning of each presentation sequence, another entry from this list is announced as the attended stimulus (see **FocusOn**). This parameter is only used in copy mode.

UserComment An arbitrary string intended for documentation purposes.

WindowBackgroundColor The window's background color, given as an RGB value. For convenience, RGB values may be entered in hexadecimal notation, e.g., 0xff0000 for red.

WindowLeft, WindowTop The screen position of the application window's top left corner, in pixels.

WindowWidth, WindowHeight The width and height of the subject-visible application window, in pixels.

10.8.2.2 States

PhaseInSequence This state is 1 during pre-sequence, 2 during sequence and 3 during post-sequence (see Timeline).

SelectedStimulus When classification is performed, this state contains the stimulus code of the stimulus classified as the "selected" one.

StimulusBegin This state is 1 during the first block of stimulus presentation, and 0 otherwise.

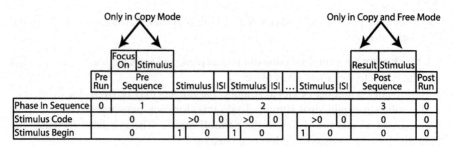

Fig. 10.13 Timeline for the Stimulus Presentation task

StimulusCode The numerical ID of the stimulus being presented (16 bit).

StimulusTime A 16-bit time stamp in the same format as the **SourceTime** state. This time stamp is set immediately after the application module has updated the stimulus/feedback display.

StimulusType This state is 1 during presentation of an attended stimulus, and 0 otherwise. The notion of an "attended" stimulus requires data recording in copy mode.

The time line of this task, as well as corresponding state values, are shown in Fig. 10.13.

10.8.2.3 Stimulus Presentation Remarks

Stimulus Definition

Stimuli are set up through a parameter defined by the application module. This implicitly defines the total number of stimuli as well as the details of each stimulus. Each stimulus is defined by the following properties:

1. Caption
2. Icon file
3. Audio file

In addition to stimuli that are part of the actual stimulation sequence, the **Focus-On** and **Result** parameters contain definitions for a stimulus that indicates what to focus on, and a stimulus that presents the result. These stimuli are only used when the task is set to copy or free mode. The following table contains an example definition of two stimuli:

	stimulus1	stimulus2
Caption	Donkey	
Icon	images/donkey.bmp	images/elephant.bmp
Audio	sounds/snicker.wav	sounds/trumpet.wav

A blank entry for caption/icon/audio file is accepted, and implies that no presentation of the respective element takes place (e.g., see "Caption" in stimulus2). The stimulus definition parameter does not contain a description on how the stimuli are presented. For further details, see the **Stimuli** parameter description.

Stimulus Codes
When defining a stimulus sequence, stimuli are referred to an integer ID called stimulus code. The stimulus code associated with a stimulus corresponds to the column in which that stimulus is defined in the **Stimuli** matrix parameter. In the recorded data file, stimulus presentation is indicated by the **StimulusCode** state. During presentation of a stimulus, this state is set to the associated stimulus code.

Stimulus Sequence
Stimuli are presented in a certain sequence. This sequence can either be deterministic, i.e., defined by the investigator, or pseudo-random.

Deterministic Sequence
The investigator defines the order by entering a list of stimulus IDs to be presented. As an example, 1 5 3 4 2 defines a sequence in which stimulus 1 is presented first, followed by stimulus 5, etc.

Random Sequence
The investigator defines absolute stimulus frequencies for each stimulus, with the sum N of those values equaling the total number of stimulus presentations in the final sequence. The resulting random sequence is obtained by applying a random permutation to an arbitrary sequence that reproduces the given frequencies, and where all $N!$ index permutations are equally probable (block randomization).

As an example, 6 2 3 defines a sequence of 11 stimulus presentations with stimulus one being presented 6 times, stimulus two 2 times, and stimulus three 3 times. The resulting sequence will be a permutation of $S_0 = [1, 1, 1, 1, 1, 1, 2, 2, 3, 3, 3]$.

Multiple sequences can be generated from the given frequencies. The investigator can define how many sequences are generated and presented.

Stimulus Delivery
For any stimulus, delivery occurs simultaneously for caption, icon, and audio. When both caption and icon are defined, the caption appears overlaying the icon.

An investigator can specify:

- Size and position of the target window.
- Width and height of caption and icon in percent of screen width/height, or that the icon should appear in its original pixel size.
- Whether captions, icons, or audio files will be presented (i.e., a global switch). There are no individual switches for each stimulus. However, individual captions/icons/wave files are not presented if they are not defined (i.e., their respective entries are blank).
- The volume for audio playback as a percentage of maximum volume.
- Window background color in RGB. (For convenience, RGB values may be entered in hexadecimal notation, e.g., 0xff0000 for red.)

- Caption color in RGB.
- The duration of stimulus presentation. (Playback of audio extending above the specified duration will be muted.)
- The duration of an inter-stimulus interval following stimulus presentation. (During the inter-stimulus interval, the screen is blank and audio is turned off.)
- Variance in inter-stimulus intervals, with probability distributed uniformly over the interval between minimum and maximum inter-stimulus interval.
- For documentation purposes, a user can enter a comment to the specific run in a string parameter.

Processing of Classification Results

The task can be configured to interpret results communicated to it by the **P3 Signal Processing** module. These results represent a judgment of which of the stimuli was most likely selected. Handling of these results is identical to the **P300 Speller**.

When it transmits a classification result, **Signal Processing** sets the state **StimulusCodeRes** to the stimulus code that was originally transmitted to it by the user application. For example, when **Signal Processing** sets **StimulusCodeRes** to 3, it transmits classification results for stimulus 3. In addition, it sets **StimulusTypeRes** to reflect the type of stimulus (0 = non-target, 1 = target) when the system is in copy mode. **Signal Processing** transmits the classification result as one number (i.e., the first control signal).

Free Mode

The task can be configured to operate in free mode. In this case, the sequence of stimulus deliveries is followed by a time period during which the Signal Processing classification result is presented. The final classification result is the stimulus with the highest classification result.

In order to present this result, the system uses the stimulus defined in the result column of the stimuli parameter. This presentation is followed by delivery of the determined stimulus. In other words, after a sequence of stimulus deliveries, the system might play a .wav file that says: "the result is," followed by a .wav file that says "yes." (assuming "yes" represents the stimulus that produced the highest classification result).

Finally, the task sends this result to the Operator module as an ASCII text message so that it appears in a log window. Free mode does not terminate until the investigator suspends operation.

Copy Mode

Copy mode is similar to free mode. In copy mode, the investigator can define a list of stimuli to be copied (e.g., 3 5 4). In this example, the user has to attend to stimulus 3 for the first sequence, 5 for the second sequence, etc.

In addition to presenting the result, the delivery of stimuli is preceded by a presentation that describes the stimulus to which the user must attend. This presentation uses the stimulus that is defined in the **FocusOn** parameter. This presentation is followed by delivery of the desired target stimulus. As an example, the system

Fig. 10.14 An exemplary 6 by 6 speller matrix. Here, the user's task is to spell the word "SEND" (one character at a time). For each character, all rows and columns in the matrix are intensified a number of times. In this example, the third row is intensified

might say "Please focus on" ... "yes," before it starts with the sequence of stimulus deliveries.

Copy mode terminates (i.e., suspends the task) when the user has finished copying all stimuli specified by the investigator.

10.8.3 P300 Speller

The **P3SpellerTask** implements Donchin's matrix speller paradigm [1, 4]. The most typical configuration is a 6 by 6 matrix (see Fig. 10.14). The user's task is to focus attention on characters in a word that is prescribed by the investigator (i.e., one character at a time) and displayed on the user's screen. Alternatively, the user may freely choose letters he/she attends to, using the speller as a true communication device. All rows and columns of this matrix are successively and randomly intensified. For a full set of intensifications of rows or columns, two contain the desired character (i.e., one particular row and one particular column). The responses evoked by these infrequent stimuli are different from those evoked by the stimuli that did not contain the desired character, in a manner similar to the classical P300 paradigm [4]. The entire set of stimuli is usually flashed several times before the **P3SpellerTask** calculates the desired character. To ensure optimal amplitudes of the evoked responses, it is important that the stimuli are infrequent even between sequences. Therefore, the **P3SpellerTask** avoids starting a sequence with the row or column that was intensified last in the previous sequence. By determining the row/column with the strongest response, the speller derives which of the matrix elements the subject wanted, and executes the action corresponding to that matrix element. Typically, this action will consist in adding a character to a text window. Beyond the basic features, the **P3SpellerTask** provides the ability to handle multiple speller menus, saving and recovering the text buffer, and an option to exchange information with an external program. Typically, the **P3SpellerTask** is used in conjunction with the **P3TemporalFilter** signal processing filter/**P3SignalProcessing** signal processing module.

10.8.3.1 Parameters

AudioStimuliOn Switches playback of audio stimuli on or off.

AudioStimuliRowsFiles, AudioStimuliColsFiles Each of these parameters is a single-column matrix specifying audio files associated with speller rows or columns, respectively. Whenever a row/column is highlighted, the associated audio file is played back. For audio files, Windows WAV format is expected. Paths may be absolute, or relative to the executable's working directory at startup, which usually matches the executable's location. Rather than the path to an audio file, text may be given, enclosed in single quotes. In this case, the text is rendered using the system's Text-To-Speech engine.

BackgroundColor Matrix elements' background color, given in RGB encoding.

DestinationAddress A network address to receive speller output, given in `IP:port` format, e.g., `localhost:3582`. A UDP socket is opened to this address; the program then sends information about selected matrix elements to the external application. For each selection, the speller will write whatever is contained in the selected matrix element's Enter field, preceded with `P3Speller_Output` and a space character, and followed with a `\r\n` sequence (i.e., an MS-DOS style line ending). E.g., the output will be `P3Speller_Output A\r\n` when a matrix element is selected that enters the letter "A," and `P3Speller_Output` `<BS>` `\r\n` for a matrix element corresponding to the backspace command.

DisplayResults Switches result display of copy/free spelling on or off.

FirstActiveMenu For multiple menus, the index of the menu that should be active at the start of a run.

IconHighlightFactor If **IconHighlightMode** is 1 or 4, this parameter defines the brightness scaling factor. Dimming is equivalent to intensifying with a scaling factor less than 1.

IconHighlightMode An enumerated value that specifies how icons are highlighted on stimulus presentation:

- 0: Show/Hide: icons are only visible during stimulus presentation,
- 1: Intensify: highlight by increasing icon brightness,
- 2: Grayscale: display a grayscale version during stimulus presentation,
- 3: Invert: invert color/brightness values during stimulus presentation,
- 4: Dim: decrease brightness during presentation.

InterpretMode An enumerated value selecting on-line classification of evoked responses:

- 0: no target is announced "attended," and no classification is performed;
- 1: online or free mode: classification is performed, but no "attended target" is defined;
- 2: copy mode: "attended" targets are defined, classification is performed.

ISIMinDuration, ISIMaxDuration Minimum and maximum duration of the inter-stimulus interval. During the inter-stimulus interval, the screen is blank, and audio is muted. Actual inter-stimulus intervals vary randomly between minimum and maximum value, with uniform probability for all intermediate values. Given in sample blocks, or in time units when immediately followed with 's,' 'ms,' or similar. Note that temporal resolution is limited to a single sample block.

NumberOfSequences The number of intensification sequences performed prior to each classification (selection of matrix elements). For an $N \times M$ speller matrix, a single intensification sequences comprises $N + M$ intensifications, one for each row, and one for each column. Usually, this parameter is set to the same value as the **P3TemporalFilter**'s **EpochsToAverage** parameter.

NumMatrixColumns, NumMatrixRows The number of columns/rows in the speller matrix. For **Multiple Menus**, a list of numbers representing each menu's number of columns/rows.

PostRunDuration Duration of the pause following last sequence. Given in sample blocks, or in time units when immediately followed with 's,' 'ms,' or similar.

PostSequenceDuration Duration of the pause following sequences (or sets of intensifications). Given in sample blocks, or in time units when immediately followed with 's,' 'ms,' or similar. When used in conjunction with the **P3TemporalFilter**, this value needs to be larger than the **EpochLength** parameter. This allows classification to complete before the next sequence of stimuli is presented.

PreRunDuration The duration of the pause preceding the first sequence. Given in sample blocks, or in time units when immediately followed with 's,' 'ms,' or similar.

PreSequenceDuration Duration of the pause preceding sequences (or sets of intensifications). Given in sample blocks, or in time units when immediately followed with 's,' 'ms,' or similar. In free or copy mode, the **PreSequenceDuration** and **PostSequenceDuration** parameters may not go below twice the value of the **StimulusDuration** parameters in order to allow for presentation of **FocusOn** and **Result** announcement stimuli.

StatusBarSize, StatusBarTextHeight The size and text height of the status bar in percent of screen height. The status bar is located on top of the screen, and displays the text that has been spelled so far. In copy mode, it also displays the text that the user is supposed to spell.

StimulusDuration For visual stimuli, the duration of stimulus presentation. For auditory stimuli, the maximum duration, i.e., playback of audio extending above the specified duration will be muted. Given in sample blocks, or in time units when immediately followed with 's,' 'ms,' or similar.

TargetDefinitions A five-column matrix defining elements of the speller matrix. Each row corresponds to a single matrix element; matrix elements are enumerated row-wise, beginning with the top left matrix element. The columns of the definition matrix are:

1. **Display:** A text string to be displayed in the matrix element, i.e., the matrix element's caption.
2. **Enter:** Specifies the speller action to be performed upon the item's selection; in most cases, this action consists in entering a text string, and is specified by that string. E.g., for the top left matrix element to display the caption "A", and also enter the letter "A" on selection, both the *Display* and *Enter* columns will contain the letter "A."

3. **Size:** Specifies an individual size for the matrix element, relative to the size of the other matrix elements.
4. **Icon File:** Contains the path to an icon file to be displayed in the speller matrix element. Windows BMP format is accepted.
5. **Sound:** Contains the path to a sound file to be played when the matrix element is selected. Sounds are given as a path to a Windows WAV file, or as a text enclosed in single quotes. When a text is given, it will be spoken using the system's Text-To-Speech engine.

TargetTextHeight Matrix elements' text height in percent of screen height.

TargetWidth, TargetHeight A single matrix element's width/height in percent of screen width/height.

TestMode If this parameter is selected, clicking on a matrix element with the mouse will select it once the current sequence of intensifications is finished. This is useful to test speller matrix configurations.

TextColor, TextColorIntensified Text color in standard and highlighted (intensified) mode, given in RGB encoding.

TextResult At the beginning of a run, this parameter's content is copied into the lower part of the status bar. At the end of a run, the status bar's content is copied back into this parameter.

TextToSpell In copy mode, a string of characters defining the text to be spelled by the user. This text is displayed in the status bar, above the actually spelled text. From the difference between **TextToSpell**, and the actually spelled text, the speller automatically derives which matrix element the user will need to select next. This information is then used to set the **StimulusType** state.

TextWindowEnabled If this parameter is selected, a separate window is displayed. Once the status bar is filled, text flows into the window, and back in case of text deletion.

TextWindowFilePath A (relative or absolute) path to a directory. Upon the <SAVE> and <RETR> speller commands, the text window's contents are saved to/retrieved from a file located in that directory. Repeated <SAVE> commands do not result in overwriting existing files. Rather, existing files are preserved, and the most recent file's name is written into a pointer file. Paths may be absolute, or relative to the executable's working directory at startup, which usually matches the executable's location.

TextWindowFontName, TextWindowFontSize Text window font name and size.

TextWindowLeft, TextWindowTop, TextWindowWidth, TextWindowHeight Position and dimension of the text window in pixels.

WindowBackgroundColor The window's background color, given as an RGB value. For convenience, RGB values may be entered in hexadecimal notation, e.g., 0xff0000 for red.

WindowLeft, WindowTop The screen position of the application window's top left corner, in pixels.

WindowWidth, WindowHeight The width and height of the subject-visible application window, in pixels.

10.8.3.2 States

PhaseInSequence This state is 1 during pre-sequence, 2 during sequence and 3 during post-sequence (see the timeline above).

SelectedTarget, SelectedRow, SelectedColumn Upon classification, these states are set to the selected target's ID, and its associated row and column, respectively. A target's ID matches its row number in the **TargetDefinitions** matrix.

StimulusBegin This state is 1 during the first block of stimulus presentation, and 0 otherwise.

StimulusCode The numerical ID of the stimulus being presented (16 bit).

StimulusTime A 16-bit time stamp in the same format as the **SourceTime** state. This time stamp is set immediately after the application module has updated the stimulus/feedback display.

StimulusType This state is 1 during presentation of an attended stimulus, and 0 otherwise. The notion of an "attended" stimulus requires data recording in copy mode.

10.8.3.3 P300 Speller Remarks

Speller Commands

Speller commands are specified in the second column of the **TargetDefinitions** matrix. Speller commands may be sequences of characters, which are added to the speller text when the respective item is selected. Additionally, speller commands may be speller control commands enclosed in a pair of $<$ $>$ characters. Any combination of characters and commands is allowed, and will be executed in sequence.

Available speller control commands are:

- $<$BS$>$ (backspace) – delete the last character from the current text.
- $<$DW$>$ (delete word) – delete the last word from the current text.
- $<$UNDO$>$ – undo the effect of the previous speller action.
- $<$END$>$ – end spelling, put BCI2000 in suspended mode.
- $<$SLEEP$>$ – suspend spelling; resume when $<$SLEEP$>$ is selected two additional times.
- $<$PAUSE$>$ – suspend spelling; resume when $<$PAUSE$>$ is selected again.
- $<$GOTO x$>$ – go to speller menu number x (see **Multiple Menus** below).
- $<$BACK$>$ – return to the previously active speller menu.
- $<$SAVE$>$ – write the text window's content into a file located at **TextWindow-FilePath**.
- $<$RETR$>$ – load the text window's content from the most recently saved file.

Multiple Menus

The **P3SpellerTask** allows you to specify a number of speller menus, and to switch between them using the $<$GOTO$>$ and $<$BACK$>$ speller commands. For multiple speller menus, the **TargetDefinitions** matrix needs to be configured as a list of

Fig. 10.15 Elements of the user's screen. The **Text To Spell** field indicates the pre-defined text. The speller will analyze evoked responses, and will append the selected text to **Text Result**

matrices rather than a single matrix. Then, each submatrix should have the form described above. Each of the submatrices may have an individual number of rows and columns and its own set of matrix elements. Also, switching to **Multiple Menus** implies additional entries in the following parameters:

- **NumMatrixColumns**, **NumMatrixRows**
- **AudioStimuliRowsFiles**, **AudioStimuliColsFiles** (additional columns)
- **TargetWidth**, **TargetHeight**, **TargetTextHeight**, **BackgroundColor**
- **TextColor**, **TextColorIntensified**, **IconHighlightMode**, **IconHighlightFactor**

Visual Representation
The visual representation is divided into three parts (Fig. 10.15):

Text to Spell displays the text that the user needs to spell (only used in copy spelling mode).
Text Result holds the letters spelled up to the current moment.
Speller Display is the area that contains the speller matrix.

In addition to the text characters, it is possible to display icons (bitmaps) in the **P3Speller** matrix. This can be achieved by entering the appropriate file name for the icon in column 4 of the desired cell in the **TargetDefinition** parameter. The icon will be flashed (highlighted) using a number of methods (see the description of the **IconHighlightMode** parameter above). As new characters are added to the right, the **Text Result** area gets filled; once there is no space left, the text is scrolled to the left in order to accommodate additional characters.

It is also possible to play a sound file or to "speak" text (using a Text-to-Speech engine) when a cell is selected. To play a sound file, the sound file name should be entered in column 5 of the desired cell in the target definition matrix. To enable a text-to-speech function, the text to be "spoken" should be entered in column 5 of the desired cell in the target definition matrix within single quotes (e.g., 'text'),

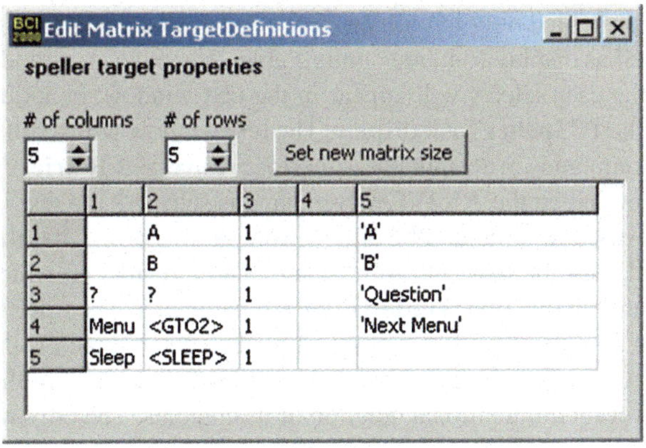

Fig. 10.16 Example for a target definition matrix

as in Fig. 10.16. The Text-to-Speech engine uses the system's "default voice" setting (*Control Panel* → *Speech* → *TextToSpeech* → *Voice Selection*). Male or female voices may be selected.

Testing Matrix Menus

In order to test matrix menu configurations, the **P3SpellerTask** may be put into a test mode. In this mode, mouse clicks into matrix elements are registered during sequence presentation, and force selection of the respective matrix element.

Pause and Sleep

There are two speller commands that allow the user to suspend operation of the **P3SpellerTask** either momentarily or for an extended period of time, <PAUSE> and <SLEEP>. To associate speller commands with certain matrix elements, specify them in the "Enter" column of the desired matrix element in the target definition matrix. The <PAUSE> speller command pauses the **P3SpellerTask**: While the system is paused, the matrix will continue to flash but target selections are ignored until the user resumes system operation, which is achieved by selecting <PAUSE> again. Data recording is also suspended while the system is paused. The **goal text** line of the status bar indicates that the system is paused. For a matrix with N entries (one of which is <PAUSE>), the chance to erroneously resume system operation while paused is $\frac{1}{N}$ per selection. On average, this will happen after N selections in paused state. A second speller command, <SLEEP>, is provided as a safer option. Once in sleep state, the system will resume only after receiving two consecutive <SLEEP> selections, which requires N^2 selections on average to occur by chance. When in sleep mode, the **goal text** line of the status bar will instruct the user to select the <SLEEP> command twice to restart.

Text Window

The capability to display user selected text in a text window is available. This feature can be activated by selecting the **TextWindowEnabled** parameter. The text window can be enabled only in online (free spelling) mode. Position and size of the window, as well as display font, are configurable. When the text window is enabled, any text that the user selects will appear in the text window, in addition to the text result area of the **P3SpellerTask** display. The text window will scroll automatically. Two speller commands are available to perform **Save** and **Retrieve** operations on the text window. When the <SAVE> command is selected, the text in the text window will be written to a file and be erased from the window. The file name will be auto generated with the date and time stamp. The directory to which the file gets sent is configured in the **TextWindowFilePath** parameter. The <RETR> (retrieve) function reads the latest file that was saved by the user and recalls the text into the text window. A separate **clear** command has not been provided in the text window functions to prevent unintentional deletion of its contents. The only way to clear the contents of the text window is using <SAVE>.

Nested Matrices

To allow for more flexibility in real-world applications, multiple speller matrices can be specified, and may be traversed in a "nested" manner. For example, one of the cells in the first matrix may be tied to another matrix which is displayed when this cell is selected by the user. This capability may be used to design a menu-based user interface.

To use the nested menu capability, the following steps need to be executed:

Set up the Target Definition Matrix To maintain backward compatibility and for ease of configuration, the application will not support nested functionality by simply converting one cell of the old target definition matrix to a sub-matrix. To define nested matrices, the **TargetDefinition** parameter must be defined with one column and each cell as a sub-matrix. If it has more than one column, the application will treat it as a single matrix and will ensure that each of its cells is defined as a single value.

Convert Each Cell to a Sub-matrix In the GUI, this is done by right-clicking on each cell as shown in Fig. 10.18, and choosing "Convert to sub-matrix" from the context menu.

Configure Individual Sub-matrices Each cell in a nested matrix is a sub-matrix (as shown in Fig. 10.17) and needs to be configured individually. In the GUI, clicking on a cell that is a sub-matrix will bring up that matrix in another window, as shown in Fig. 10.19. Each sub-matrix should be configured to have a minimum of three columns and can have up to five columns if it needs to display icons or play sounds.

Enable Transition from One Menu to Another Special control codes (speller commands) allow transition from one matrix to another. To go to a different matrix when an item is selected, enter the <GOTO#> command into its "Enter" column, replacing "#" with the (one-based) index of the corresponding matrix (menu). The <BACK> (or <BK>) command will go back to the previous menu/matrix (see Fig. 10.19).

Number of Rows and Columns of Each Nested Menu (Matrix) The number of rows and columns in each sub-matrix needs to be entered into the **NumMa-**

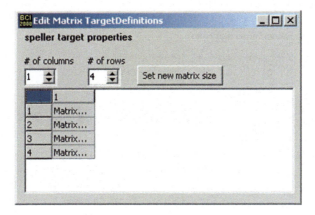

Fig. 10.17   Example for a target definition matrix for four nested menus

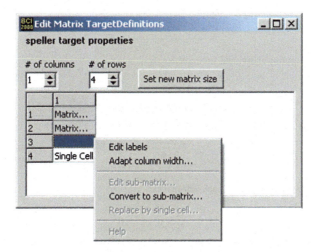

Fig. 10.18   Converting cells to sub-matrices

trixRows and **NumMatrixColumns** parameters. There, individual entries are separated by space characters. Figure 10.20 indicates that the first three sub-matrices are 2×2, while the sub-matrix at index 4 is a 6×6 matrix. In case of a single (non-nested) matrix, there will be only one entry in the **NumMatrixRows** and **NumMatrixColumns** parameters.

Select First Menu to Be Displayed   The index of the menu to be displayed first (i.e., when the application is started) is entered in the **FirstActiveMenu** parameter. In case of non-nested matrix configuration, this parameter should be left at its default value of 1.

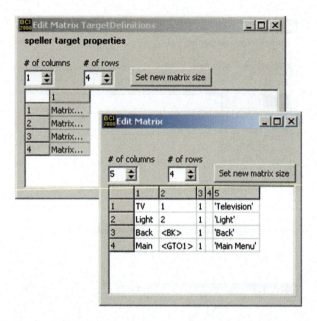

Fig. 10.19 Configuring individual sub-matrices

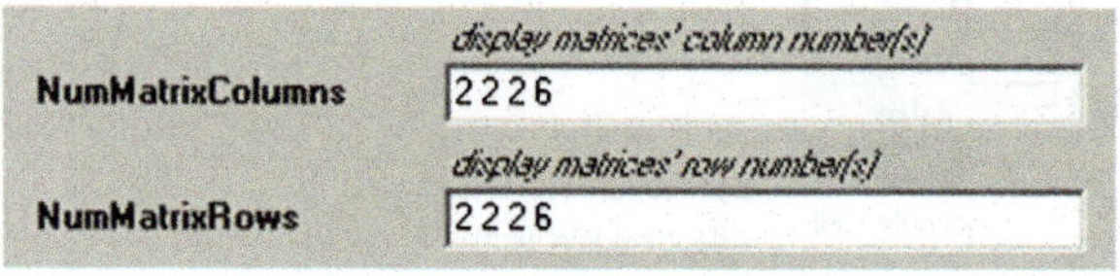

	display matrices' column number[s]
NumMatrixColumns	2 2 2 6
	display matrices' row number[s]
NumMatrixRows	2 2 2 6

Fig. 10.20 Rows and columns of each sub-matrix

10.8.4 Keystroke Filter

The **KeystrokeFilter** translates BCI2000 state changes into simulated key presses. Simulated keypresses can then be used to control an external application program installed on the machine running the BCI2000 application module. Only the lower 4 bits of the state value are used, and translated into presses of the keys corresponding to their hexadecimal representation ($0 \ldots 9$, $A \ldots F$).

10.8.4.1 Parameters

KeystrokeStateName The name of a BCI2000 state to be converted into simulated keystrokes.

10.8.4.2 States

Any existing state may be given in the **KeystrokeStateName** parameter.

10.8.5 Connector Filters

The **Connector Filters** provide an implementation of the **BCI2000 App Connector protocol**. Typically, the **ConnectorInput** filter is the first filter in the Application Module, which allows external changes to BCI2000 states to affect the application module's behavior immediately. Similarly, the **ConnectorOutput** filter is placed last in the application module, such that state information read over the protocol will immediately reflect the application module's state changes. A description of the connector protocol, along with client example code, and configuration examples, is detailed in the App Connector section in Sect. 6.4.

10.8.5.1 Parameters

ConnectorInputAddress An `address:port` combination that specifies a local UDP socket. Addresses may be host names or numerical IP addresses. Incoming data is read from this socket.

ConnectorInputFilter For incoming values, messages are filtered by state name using a list of allowed names present in this parameter. To allow signal messages, allowed signal elements must be specified including their indices. To allow all names, enter an asterisk (*) as the only list entry.

ConnectorOutputAddress An `address:port` combination that specifies a local or remote UDP socket. Addresses may be host names, or numerical IP addresses. For each data block, values of all state variables and the control signal are written out to this socket, using the App Connector protocol.

10.8.5.2 States

All states present in the system are transmitted over the App Connector protocol.

10.9 Tools

10.9.1 BCI2000 Offline Analysis

"BCI2000 Offline Analysis" is a simple tool to analyze BCI2000 data in the time- or frequency domain. To run this tool, whose screen shot is shown in Fig. 10.21, you'll need a system running either Windows 2000 or later. If your system meets this requirement, please choose from one of the following:

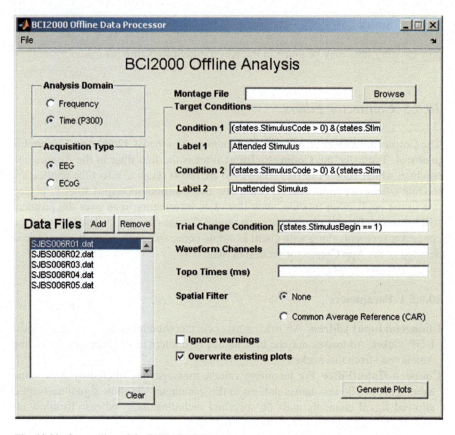

Fig. 10.21 Screen shot of the BCI2000 Offline Analysis tool

10.9.1.1 Systems with Matlab v7.0 or Greater Installed

Startup Script

If you have not yet opened Matlab, the easiest way to run BCI2000 Offline Analysis is to double click the OfflineAnalysis.bat file that resides in BCI2000/tools/OfflineAnalysis/.

From the Matlab Command Line

If MATLAB is already running, navigate to BCI2000/tools/OfflineAnalysis/ using the "Current Directory" explorer, or type the absolute path into the textbox at the top of the Matlab window. Then, type OfflineAnalysis at the MATLAB command prompt to start analyzing your data.

10.9.1.2 Systems That Do Not Have Matlab Installed or with Versions That Predate MATLAB v7.0

Offline Analysis Without Matlab

1. Download the Matlab Component Runtime (MCR) from http://www.bci2000.org/downloads/bin/MCRInstaller.exe.
2. When the download is complete, run the installer.
3. Follow the on-screen instructions to complete the installation.
4. Navigate to `BCI2000/tools/OfflineAnalysis/`.
5. Double-click `OfflineAnalysisWin.exe` to begin analyzing your data.

10.9.1.3 Reference

File>Save Settings "Save Settings" allows you to save the values you've entered for all fields excluding "Data Files," "Ignore warnings" and "Overwrite existing plots." This functionality may be particularly useful if you want to run the same or similar analyses between sessions.

File>Load Settings "Load Settings" allows you to load a settings file previously created using File>Save Settings.

Analysis Domain The BCI2000 Offline Analysis tool allows you to analyze your data either in the frequency domain or the time domain. Be aware that certain fields will be labeled differently depending on the domain that is chosen. Specifically, choosing "Frequency" will make the "Spectra Channels" and "Topo Frequencies" fields available. Choosing "Time" will make the fields "Waveform Channels" and "Topo Times" available. Additionally, the values for the target conditions, trial change condition and spatial filter will be overwritten so as to correspond with the chosen domain.

A frequency domain analysis will result in up to three plots comprising a feature map, a spectra plot (one spectrum for each channel specified) and up to nine topographic plots (one for each topo frequency specified). For EEG-based analyses, the best frequency domain features are typically found between 9–12 Hz or 18–25 Hz. Please see the user tutorial in Sect. 5.2.3 for additional information.

A time-domain analysis will result in up to three plots that consist of a feature map, a waveform plot (consisting of one waveform for each channel specified) and up to nine topographic plots (one for each topo time specified). Typically, P300 features are found at around 300 ms following presentation of the stimulus. Please see the user tutorial in Sect. 5.3.3 for additional information.

Acquisition Type Acquisition type allows the user to indicate whether the data was acquired using EEG or an ECoG.

Data Files Any number of BCI2000 data files with the same state information and number of channels may be analyzed in a single run. To select a file to add to the list of files to be analyzed, simply click the "Add" button, choose the desired file and click "Open." To add multiple files from the same directory, click on any one of the desired files and then control-click the additional files. Then, when all desired files are selected, click "Open." If files have already been added to the data files list, then subsequent uses of the above procedure will append the selected files to the list allowing for addition of files from different directories.

Montage File (optional) If topographic plots are desired, you must specify a montage file that is congruent with the chosen acquisition type (i.e., EEG or ECoG). If you do not yet have a montage file, you can create one using the "Eloc Helper" tool that comes with BCI2000 Offline Analysis. In place of a montage file, you may also use data files that specify channel names that are compliant with the 10–20 standard. In this case, "BCI2000 Offline Analysis" will use the channel name corresponding to each electrode to infer the location of that electrode.

Target Conditions Target conditions consists of four fields that allow for the specification of two different conditions. Each condition consists of the condition itself (i.e., a Matlab-syntax boolean statement) and the label for the corresponding data (e.g., "Opening and Closing Both Hands"). Typically, you will want to specify two different conditions in order to compare two different datasets. Occasionally, however, you may want to inspect a single dataset to investigate artifacts. In this case, you can omit the second condition (i.e., both the condition and the label).

Trial Change Condition The trial change condition allows for the user to specify which condition indicates a change in trials. The condition must be a Matlab-syntax boolean statement.

Spectra Channels (optional) Note: This field is available only if you've chosen "Frequency" as the analysis domain. If you would like to have the analysis generate spectra plots, you must enter at least one channel in this field. Multiple channels can be specified using a comma-delimited list (e.g., "1,2,3") or a space-delimited list (e.g., "1 2 3").

Topo Frequencies (optional) Note: This field is available only if you've chosen "Frequency" as the analysis domain. If you would like to have the analysis generate topographic plots, you must enter at least one frequency in this field. Multiple frequencies can be specified using a comma-delimited list (e.g., "8,11,22") or a space-delimited list (e.g., "8 11 22").

Waveform Channels (optional) Note: This field is available only if you've chosen "Time" as the analysis domain. If you would like to have the analysis generate P300 waveform plots, you must enter at least one channel in this field. Multiple channels can be specified using a comma-delimited list (e.g., "1,2,3") or a space-delimited list (e.g., "1 2 3").

Topo Time (optional) Note: This field is available only if you've chosen "Time" as the analysis domain. If you would like to have the analysis generate topographic plots, you must enter at least one time in this field. Multiple times can be specified using a comma-delimited list (e.g., "200,300,400") or a space-delimited list (e.g., "200 300 400").

Spatial Filter By applying the right spatial filter to your data, you will likely be able to achieve better results. Depending on the data, however, the optimal filter will differ. "BCI2000 Offline Analysis" offers the common average reference

(CAR) filter, which is often useful if the response isn't spread out over a large number of electrodes. To apply a CAR filter to your data before processing, choose "Common Average Reference." Otherwise, choose "None" in order to run the analysis on the raw data.

Ignore warnings In order to get an accurate representation of the data, it is recommended that the analysis be performed on no less than 10 trials. "BCI2000 Offline Analysis" will generate a warning if this condition is not met. This warning can be ignored by checking this box. If a particular combination of data files and conditions results in less than three trials, an error is displayed and the analysis is halted. Errors cannot be overridden.

Overwrite existing plots Each time the analysis is run, up to three plots may be generated. If you want to compare the results from different runs, ensure that this checkbox is unchecked. In this case, "BCI2000 Offline Analysis" will generate the plots in new figures instead of simply overwriting the plots from previous analyses.

10.9.1.4 Troubleshooting

1. **Clicking on any button results in the error "Undefined command/function ..." followed by "Error while evaluating figure ..."** If you are receiving this error, it is possible that after opening "BCI2000 Offline Analysis" you changed Matlab's working directory. Make sure that your working directory is the same directory that contains the OfflineAnalysis script.
2. **The analysis doesn't seem to be generating any topographic plots.** In order to generate topographic plots, it is necessary to specify a montage file and at least one topo frequency (for frequency domain analyses) or topo time (for time domain analyses). If the analysis you are running doesn't generate a topographic plot, please ensure that you've entered values in both of these fields.
3. **The analysis doesn't seem to be generating any spectra or waveform plots.** To generate spectra or waveform plots, it is necessary to specify at least one spectra channel (for frequency domain analyses) or waveform channel (for time domain analyses). If your analysis is not resulting in one of these plots, please ensure that you have entered values into the appropriate field.
4. **I've typed in a valid target or trial change condition, but I get the error "Invalid ... condition specified."** Condition syntax is case-sensitive. For instance, while states.TargetCode may be a valid state condition, states.targetcode is not. Please double-check your conditions to ensure that the state variables are capitalized correctly.

10.9.2 USBampGetInfo

This command line tool displays all connected g.USBamps, including their serial number and the USB port that they connect to. Further, this tool reads all supported bandpass and notch filter configurations. Thus, this tool can be used to determine which filters can be used for a particular sampling frequency within BCI2000. An example screen output can be seen in Appendix A.

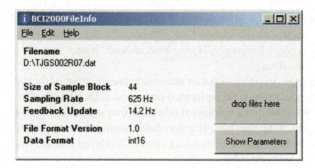

Fig. 10.22 The BCI2000FileInfo tool

10.9.3 BCI2000FileInfo

BCI2000FileInfo allows you to display and extract information from a BCI2000 data file. Its main window, which is shown in Fig. 10.22, displays information about a file's binary data format, sampling rate, sample block size, and update rate.

10.9.3.1 Viewing and Saving Parameters

Clicking the *Show Parameters* button opens up a parameter editor similar to the one provided by the Operator module. From there, all parameters or a subset of all parameters may be saved into BCI2000 parameter files.

10.9.3.2 Opening Files

In addition to the *File → Open...* menu item, files may opened by dragging them to the *Drop Files Here* area, or onto the program icon.

10.9.4 BCI2000Export

BCI2000Export (see Fig. 10.23) is a drag-and-drop-based program that allows you to import BCI2000 *.dat files into BrainProducts' VisionAnalyzer program, and to convert BCI2000 *.dat files into ASCII files.

10.9.4.1 General Usage

You may learn about the program's options by starting it and using the help hints that appear when stopping the mouse inside one of its four main areas.

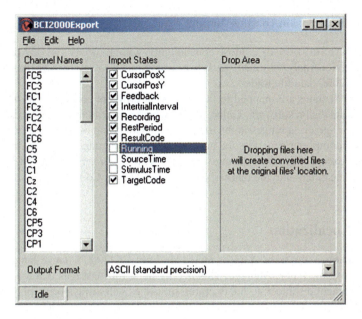

Fig. 10.23 The BCI2000Export tool

Output Format
Choose between space-delimited, tabular ASCII output and BrainVisionAnalyzer Import formats.

Importing Files
To convert files, drop them onto the "drop area" located inside its main window, or onto its application icon. Previous settings apply; BCI2000 states not found in the "import states" list will be added to the list and imported.

Excluding States
To populate the list of import states, drop BCI2000 files onto the "import states" list. You can suppress the import of a state by un-checking its name in the list.

Channel Names
Channel names are given as a list of strings in the order in which the respective channels appear in the file.

10.9.4.2 ASCII Export

An ASCII file is a space-delimited matrix with columns corresponding to channels and states, and rows corresponding to samples. The first row is a list of column headers.

10.9.4.3 BrainVisionAnalyzer Import

To use `BCI2000Export` most conveniently with `BrainVisionAnalyzer`, use the VisionAnalyzer's *Edit Workspace...* command to set the raw files folder to your BCI2000 data file folder.

BCI2000 states are mapped to the somewhat different concept of *Markers* used by the BrainVision software. Unlike states, which are written once for each sample, markers are given as ranges in the time domain, extending over arbitrary time intervals. In the importer program, the basic idea is to create one marker object for each state run except those runs during which the state is zero. For multi-bit states, the state's value enters into the marker's name, e.g., "Target Code 2."

10.10 Localization

Localization is the translation of messages from BCI2000's native language, English, into some other language. This is only done for messages that are visible to the experimental subject; to the operator, messages will always be displayed in English. Currently, localization support is limited to languages that allow for a standard 8-bit character representation, i.e., Western languages.

Localization is applied by matching strings against a translation table, and replacing with a localized version if available.

LocalizedStrings A matrix parameter that defines string translations. In this matrix, column labels are strings in their native (English) version; row labels are language names. When a string is to be translated, it is matched against column labels, and its translation will be taken from the row corresponding to the target language. Strings that do not appear as a column label will not be translated. Also, strings with an empty translation entry in **LocalizedStrings** will not be translated. Adding translations into another language is simple, and consists in

- adding a row to the **LocalizedStrings** matrix,
- labeling that row with the name of the target language,
- entering translations for the strings appearing as column titles.

Language This parameter defines the language into which strings are translated; if its value matches one of the **LocalizedStrings** row labels, translations will be taken from that row; otherwise, strings will not be translated. A value of `Default` results in all strings keeping their original values.

10.11 P300 Classifier

10.11.1 Introduction

The P300 Classifier is a tool that builds and tests a linear classifier for detection of evoked potentials (e.g., the P300 response) in data collected with BCI2000. It currently supports the P3Speller and Stimulus Presentation paradigms. The program generates the linear classifier using several methods and algorithms, most notably Stepwise Linear Discriminant Analysis (SWLDA) [2, 3, 5, 6]. The classifier derived by this program can be saved and imported into BCI2000 as a parameter file fragment (*.prm) to be used in online testing. The most common use for this program is to optimize spelling performance when using the P300 Speller module.

Unlike the Matlab-based P300 GUI that was provided with earlier version of BCI2000, the P300 Classifier is a standalone executable that does not depend on Matlab. Its core functionality is written in C++ – its graphical user interface is written using the platform independent toolkit Qt. In addition to its GUI-based functionality, the P300 Classifier program is completely scriptable, i.e., the user can execute the program using the command line and parameterize it using command line parameters. The P300 Classifier mainly realizes two functions. First, it can build a classifier using data collected using the BCI2000 P3Speller or Stimulus Presentation paradigms. Second, it can apply this classifier to data collected using these paradigms to determine classifier performance.

10.11.2 Interface

10.11.2.1 Data Pane

The P300 Classifier GUI is composed of three panes: Data, Parameters, and Details. The Data Pane shown in Fig. 10.24 allows the user to: load training and testing data files and an INI file; generate and apply feature weights; and write a parameter file fragment for use in online testing.

- **Load Training Data Files:** Use this button to load BCI2000 data files for classifier training. The information for the selected files will appear at the top of the button.
- **Load Testing Data Files:** Use this button to load BCI2000 data files for classifier testing. The information for the selected files will appear at the top of the button. Training and testing data files must be compatible.
- **Load Ini File:** Use this button to load an INI file that includes all classifier parameters:

```
[Initialization]

maxiter = 60 // maximum # features
penter = 0.1000 // probability for entering feature
premove = 0.1500 // probability for removing feature
spatial_filter = 1 // Spatial filter (1=RAW; 2=CAR)
decimation_frequency_Hz = 20 // decimation freq. in Hz
channel_set = 1 2 3 4 5 6 7 8 // select channel subset
Resp_window_ms = 0 800 // response window in ms
```

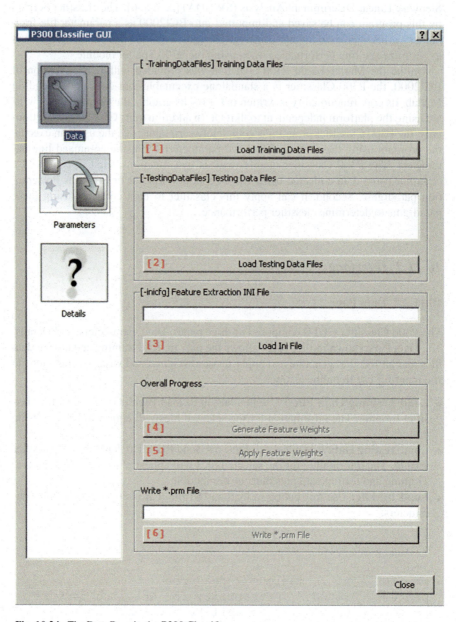

Fig. 10.24 The Data Pane in the P300 Classifier

- **Generate Feature Weights:** Use this button to generate the linear model, i.e., the weights for the different features (i.e., signal amplitudes at different times and channels), after properly configuring all of the parameters in the Parameters Pane. This button is enabled only if the parameters are properly configured and the training data files exist. Once the feature weights have been computed, a suggested name for the parameter file fragment (*.prm) will show up at the top of the "Write *.prm File" button.
- **Apply Feature Weights:** Use this button to test the classification accuracy of the feature weights currently stored in the program. The classification results will appear in the Details Pane.
- **Write *.prm File:** Use this button to save the parameter file fragment with the name suggested at the top of this button. This fragment can be loaded into BCI2000 for online testing of the feature weights.

10.11.2.2 Parameters Pane

The Parameters Pane shown in Fig. 10.25 contains all the parameters needed to generate feature weights by using the SWLDA algorithm. These parameters can be loaded using the "Load Ini File" button in the Data Pane. If the parameters are properly configured, the "Generate Feature Weights" button is enabled in the Data Pane.

- **Max Model Features:** Used to specify the maximum number of features to be used in the SWLDA algorithm. Only a single value can be entered for evaluation. The default value is 60.
- **Penter:** Used to specify the maximum p-value for a variable to be included in the model. The default value is 0.1. Penter must be less than Premove and $0 < $ Penter < 1. Only a single value can be entered for evaluation.
- **Premove:** Used to specify the maximum p-value for a variable to be removed from the model. The default value is 0.15. Premove must be greater than Penter and $0 < $ Premove < 1. Only a single value can be entered for evaluation.
- **Spatial Filter:** Selects the spatial filter applied to the training data. Select "RAW" or "CAR" from the drop-down menu. "RAW" means that there is no spatial filter applied to the data, whereas "CAR" realizes a common average reference filter using all of the channels contained in the data file, i.e., not just the channels specified in the channel set. The default spatial filter is "RAW."
- **Decimation Frequency:** Used to specify the temporal decimation frequency of the data in Hz. Only a single value can be entered for evaluation. Set this parameter to the sampling rate used in the data file for no decimation. The lower the Decimation Frequency, the smaller the feature space.
- **Channel Set:** Used to specify the channel set that will be used to create feature weights. The specified channels must be a subset of the channels contained in the training data file.
- **Response Window:** Used to specify the beginning and end time (defined in ms) following the stimuli that will be used in the analysis. These two values are automatically converted into samples according to the data's sampling rate. Only

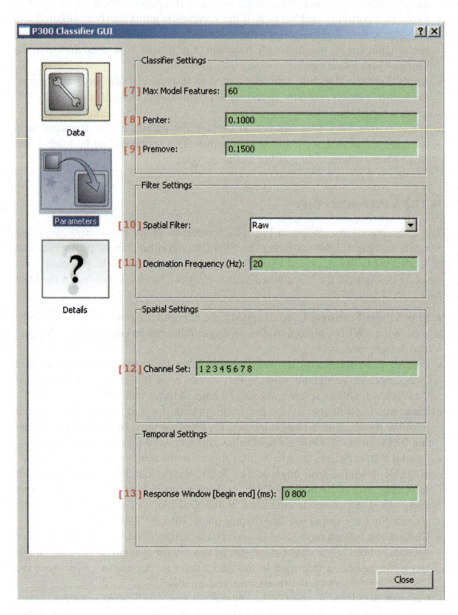

Fig. 10.25 The Parameters Pane in the P300 Classifier

a single data window can be entered and will be evaluated. The beginning time must be less than the end time.

10.11.2.3 Details Pane

The Details Pane (see Fig. 10.26) displays information about the selected BCI2000 training data files and the classification results.

- **Sampling Rate:** Displays the sampling rate of the training data files in Hz.
- **Number of Channels:** Displays the total number of channels contained in the training data files.
- **Classifier:** Displays the classifier algorithm applied to generate the feature weights. The P300Classifier currently only supports the SWLDA classifier algorithm.
- **Application:** Displays the type of paradigm. Currently, "P3SpellerTask" and "StimulusPresentationTask" are supported.
- **Interpret Mode:** Displays the interpretation mode used in the paradigm. The mode can be either "Copy Mode" or "Online Free Mode."
- **Duration:** Displays the duration of all training data files in seconds.

The text display shows the classification results for the training and testing data files.

10.11.3 Reference

The P300Classifier trains and tests a linear classifier (SWLDA) to detect evoked potentials in brain signals. Classifier training consists of the following steps:

- Load BCI2000 data files.
- Get P300 responses.
- Generate feature weights for a linear, model using SWLDA.
- Apply linear classifier to get scores.
- Interpret scores according to the given application.

The testing consists of the same steps, except that the P300 Classifier will use the linear model derived using the training dataset. These steps are described in more detail below.

- **Step 1: Load BCI2000 Data Files.** The first step to using the P300Classifier is to load training and testing data files. These BCI2000 data files are checked for compatibility and consistency. If all training data files are valid, then the "Generate Feature Weights" button is enabled and each file is colored with light green. Otherwise, the "Generate Feature Weights" button is not enabled, and each file is colored either with yellow or pink. Thus, green, yellow, and pink colors mean that the respective file is valid, may be valid but there is a mismatch with another file, or is invalid, respectively.

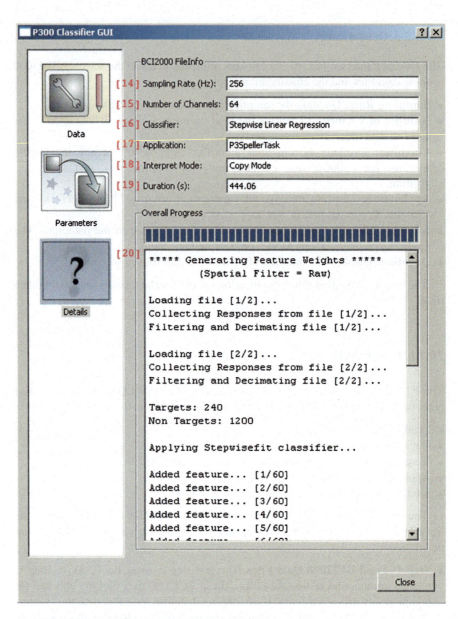

Fig. 10.26 The Details Pane in the P300 Classifier

- **Step 2: Get P300 Responses.** Signals, states, and parameters are extracted from BCI2000 training and testing data files. The features used for classification are then extracted from the brain signals using the "Response Window" and "Channel Set" specified in the Parameters Pane. These signals are then filtered and downsampled. The filter employed here is a Moving Average (MA) Filter implemented as "Direct form II Transposed." The MA filter belongs to the finite impulse response (FIR) filter category and works as a low pass filter to remove high frequency components of the signal.
- **Step 3: Generate Feature Weights for a Linear Model Using Stepwise Linear Discriminant Analysis (SWLDA).** The SWLDA algorithm derives a linear model that relates a set of input features (i.e., filtered brain response) to output labels (i.e., whether or not a P300 response is expected in that brain response) using multiple linear regressions and iterative statistical procedures. Thus, this procedure selects only those brain signal features that are important for distinguishing responses with from responses without P300 responses.
- **Step 4: Apply Linear Classifier to Get Score.** Scores are computed based on the variables included in the final linear model and the corresponding feature weights.
- **Step 5: Interpret Scores.** Scores are interpreted according to the given application. The P300 Classifier currently supports either the BCI2000 P300 Speller or StimulusPresentation modules.

10.11.4 Tutorial

Steps to Generate Feature Weights

- Press the "Load Training Data Files" button in the Data Pane.
- From the dialog box, select the desired BCI2000 *.dat file(s) for training. Selected files can be from different sessions of the same paradigm, but must contain consistent parameters. Each training data file is colored according to the color coded scheme explained earlier. Files can only be selected from a single directory; thus, the desired training data files should be organized into the same directory prior to using the P300Classifier.
- Once the Parameters Pane is correctly configured, the "Generate Feature Weights" button is enabled. Press this button to perform the analysis and to generate feature weights. The "Overall Progress" bar displayed in Data and Details panes will indicate the progress of the SWLDA. The classification results will appear in the text window of the "Details" pane. Every time you press the "Generate Feature Weights" button, and the analysis is successfully completed, there will be a new set of feature weights.
- The training procedure can be repeated multiple times with different parameters.

Steps to Generate Feature Weights (continued)

- After training is completed and feature weights are generated, it is recommended to test (cross validate) the feature weights on independent data (i.e., testing data files) before saving the parameter file fragment (*.prm). Additionally, there is a suggested parameter file fragment name that is displayed on the top of "Write *.prm File" button, which can be used to save the parameter file either with the suggested name or the name specified by the investigator.

Steps to Test the SWLDA Classifier

The Data Pane contains the "Apply Feature Weights" button to test the generated feature weights on one or more BCI2000 testing data files. The investigator must follow the next steps for applying feature weights to testing data files:

- Once feature weights are generated, and testing data files exist and have been validated, the "Apply Feature Weights" button is enabled. If no testing data files are loaded, the investigator must press this button to load testing data file(s). One or more BCI2000 *.dat files can be selected from the same directory as a single "test file group" each time the button is pressed. Selected files can be from different sessions of the same paradigm, but must contain consistent parameters. Each testing data file is colored according to the color coded scheme explained earlier. All selected test files should have the same sampling rate and electrode montage as the training data files that were used to generate the current feature weights set.
- After all testing data files are selected, press the "Apply Feature Weights" button to perform the analysis. Results of the classification are displayed in the text window of the Details Pane. The "Overall Progress" bar displayed in the Data and Details panes will indicate the progress of the classification.
- After evaluating the classification results, the *.prm file from the current session can be saved by pressing the "Write *.prm File" button.

10.11.5 Example

In this example, we are interested in computing feature weights from the BCI2000 training data file eeg3_1.dat that was created using the P300 Speller paradigm and is provided with the BCI2000 distribution.

First, press the "Load Training Data Files" button. Notice that the training data file is set up correctly since it is colored in light green. Remember that the classifier algorithm will be trained using this training data file.

The parameters shown in the Parameters Pane are set by default. For illustration purposes, we may create an INI file with the following content and then load that file using the "Load Ini File" button.

```
[Initialization]
maxiter = 60
penter = 0.1000
premove = 0.1500
spatial_filter = 2
decimation_frequency_Hz = 20
channel_set = 1 2 3 4 5 6 7 8
Resp_window_ms = 0 800
```

Once the initial parameters are loaded and properly set up, the "Generate Feature Weights" button is enabled, and each one of the parameter fields is colored in light green. The investigator can change any of the parameter fields at any time. However, if the given parameter is invalid, the corresponding parameter field will turn pink. Once the parameters are properly configured, the investigator may generate feature weights by pressing the "Generate Feature Weights" button. Details of the training data file and progress of the classification are displayed in the Details Pane. Once the classifier has been trained with the BCI2000 training data file, the next step is to test the derived feature weights with the BCI2000 testing data file eeg3_2.dat. By pressing the "Load Testing Data Files" button, this file is loaded into the P300 Classifier program. Notice that the testing data file is set up correctly since it is colored in light green. Because the testing data file is valid, the "Apply Feature Weights" button is enabled. Press the "Apply Feature Weights" button. Details of the training data file (no testing data file) and progress of the classification are displayed in the Details Pane. You can now write a parameter file fragment *.prm with the name suggested in the "Write *.prm File" field by pressing the "Write *.prm File" button. This parameter file fragment can be imported into BCI2000 for online testing of the feature weights. Remark: every time the investigator changes any of the parameters in the Parameters Pane, the "Apply Feature Weights" button is disabled. To enable it, the investigator must generate new feature weights.

References

1. Donchin, E., Spencer, K.M., Wijesinghe, R.: The mental prosthesis: assessing the speed of a P300-based brain–computer interface. IEEE Trans. Rehabil. Eng. **8**(2), 174–179 (2000)
2. Draper, N., Smith, H.: Applied Regression Analysis. Wiley–Interscience, New York (1998)
3. Embree, P., Kimball, B.: C Language Algorithms for Digital Signal Processing. Prentice–Hall, New York (1991)
4. Farwell, L.A., Donchin, E.: Talking off the top of your head: toward a mental prosthesis utilizing event-related brain potentials. Electroencephalogr. Clin. Neurophysiol. **70**(6), 510–523 (1988)
5. Press, W., Flannery, B., Teukolsky, S., Vetterling, W.: Numerical Recipes in C: The Art of Scientific Computing. Cambridge University Press, Cambridge (1992)
6. Ralson, A., Wilf, H.: Mathematical Methods for Digital Computers. Wiley, New York (1967)
7. Taylor, D.M., Tillery, S.I., Schwartz, A.B.: Direct cortical control of 3D neuroprosthetic devices. Science **296**, 1829–1832 (2002)
8. Wolpaw, J.R., McFarland, D.J.: Control of a two-dimensional movement signal by a noninvasive brain–computer interface in humans. Proc. Natl. Acad. Sci. USA **101**(51), 17849–17854 (2004). doi:10.1073/pnas.0403504101

Chapter 11
Contributed Modules

Any BCI2000 distribution consists of a *core* and a *contribution* part. Core components (i.e., modules, or filters) are maintained by the BCI2000 team. We make every attempt to optimize the quality of these components and their documentation. Contributions are usually components that have been created by the BCI2000 community. Because we often cannot replicate the environment that is necessary for execution of these components (this is true particularly for contributed Source modules that make use of other data acquisition devices), we cannot enforce the same level of quality as for the core components. The following sections describe these contributions.

11.1 Source Modules

11.1.1 Amp Server Pro

The **AmpServerProADC** component supports devices by Electrical Geodesics Incorporated (EGI) by implementing the client side of EGI's TCP/IP-based Amp Server Pro (ASP) protocol. Thus, it may be used to interface BCI2000 with an EGI amplifier managed by an ASP server.

11.1.1.1 Authors

Joshua Fialkoff, Wadsworth Center, New York State Department of Health, 2008.

11.1.1.2 Using the Amp Server Pro Source Module

Amp Server Pro is capable of working with many amplifiers concurrently. To begin using Amp Server Pro with BCI2000, ensure that at least one amplifier is connected to the server. If no amplifier is connected, the Amp Server software emulates an amplifier. If you choose to use the emulated amplifier, you should expect to see a smooth sine wave signal for all channels. To start the amp server, simply navigate to the Amp Server Pro directory and double click the file named `Amp Server`.

11.1.1.3 Compiling the Amp Server Pro Source Module

Because the Amp Server Pro Source Module is a contributed module, you either need to download the binary distribution for the contributed components, or must compile the module from the source code before it can be used. To compile the Amp Server Pro module, you need Borland C++ Builder v6.0 or later. To proceed, please follow the instructions below:

1. Navigate to `BCI2000/src/contrib/SignalSource/AmpServerPro`.
2. Double click `AmpServerPro.bpr` to open the `Amp Server Pro project file` in the Borland C++ Builder.
3. From the file menu click *Project* → *Make AmpServerPro*.

11.1.1.4 Parameters

AmplifierID Amp Server Pro is capable of managing many amplifiers concurrently. BCI2000 operates with one of these amplifiers. If only a single amplifier is connected, you may enter the value "auto" and allow BCI2000 to automatically determine the Amplifier ID. If multiple amplifiers are connected, you must enter a valid ID from 0 to $N - 1$ where N is the number of amplifiers connected.

CommandPort The port number used for command layer communication. Unless you have explicitly set the port number via Amp Server's configuration, the default of 9,877 should be correct.

NotificationPort The port number used for notification layer communication. Unless you have explicitly set the port number via Amp Server's configuration, the default of 9,878 should be correct.

ServerIP The IP address of the computer running the Amp Server software (e.g., 192.168.0.3).

StreamPort The port number of the port used for data streaming. Unless you have explicitly set the port number via Amp Server's configuration, the default of 9,879 should be correct.

11.1.1.5 States

None.

11.1.2 BioRadio

This component supports the BioRadio150 EEG amplifier produced by Cleveland Medical Devices.

11.1.2.1 Authors

Yvan Pearson-Lecours.

11.1.2.2 Installation

Make sure that the `BioRadio150DLL.dll` file is available in the directory that contains the BioRadio Source module.

11.1.2.3 Parameters

COMPort BioRadio150 COM port. Zero sets the port to AUTO, and a number between one and fifteen sets it to the corresponding COM port (e.g., 1 sets to COM1, 2 to COM2, etc.).
ConfigPath Absolute path to the BioRadio150 configuration file.
VoltageRange BioRadio150 Voltage Range:

- $7 = \pm 100$ mV,
- $6 = \pm 450$ mV,
- $5 = \pm 25$ mV,
- $4 = \pm 412$ mV,
- $3 = \pm 6$ mV,
- $2 = \pm 3$ mV,
- $1 = \pm 1.5$ mV,
- $0 = \pm 750$ μV.

11.1.2.4 States

None.

11.1.3 BioSemi 2

This Source module supports acquisition from Biosemi devices.

11.1.3.1 Authors

Samuel A. Inverso (samuel.inverso@gmail.com), Yang Zhen, Maria Laura Blefari, Jeremy Hill and Gerwin Schalk.

11.1.3.2 Installation

Copy the `Labview_DLL.dll` to the prog directory for the module to work.

11.1.3.3 Parameters

As of version 2.0 (svn revision 2189), the old parameters **PostFixTriggers** and **TriggerScaleMultiplier** have been removed. The functionality of the former is replaced by **TriggerChList** and the latter can be simulated by setting the corresponding elements of **SourceChGain** accordingly.

AIBChList A list of indices (each in the range 1 through 32 inclusive) that indicate which Analog Input Box channels are to be acquired. By default, none are acquired. AIB channels are postfixed immediately following the EEG channels.

EEGChlist A list of (one-based) indices to the EEG channels to be acquired. If there are n indices in this list, then the first n channels will be EEG channels.

TriggerChList The old PostfixTriggers parameter has been replaced by this parameter, which gives a list of indices (each in the range 1 through 16 inclusive) indicating which one-bit trigger channels to postfix after the EEG and AIB channels. By default, all 16 are postfixed.

11.1.3.4 States

BatteryLow Set to 1 when hardware reports low battery state.
MK2 Set to 1 when connected to an MK2.
MODE The mode corresponding to the one on the front of the Biosemi box.

11.1.3.5 Known Issues

Pressing *Set Config* more than once per BCI2000 launch currently (revision 2190) results in unstable amp behavior.

11.1.4 Measurement Computing

The DAS_ADC component handles A/D boards from Measurement Computing (previously called ComputerBoards).

This ADC has been tested and proven to work on the following boards as displayed by the InstaCal program:

DEMO-BOARD
PC-CARD-DAS16/16
PCM-DAS16S/16
CIO-DAS1402/16
PCIM-1602/16

11.1.4.1 Authors

Jürgen Mellinger (juergen.mellinger@uni-tuebingen.de).

11.1.4.2 Installation

The DAS source module uses Measurement Computing's system wide driver
and configuration files. If there is an error message saying that loading the
DLL failed, you need to download and install a recent version of InstaCal from
http://www.measurementcomputing.com (free of charge), and then use it to configure your board.

For the source module to work properly, you need to delete any files named
cbw32.dll and cb.cfg from the directory that contains the Source module. It is
also a good idea to delete all other files called cbw32.dll and cb.cfg, except
the ones located in the directory where Measurement Computing's InstaCal.exe
resides (usually C:/mmc or C:/Program Files/mmc).

11.1.4.3 Parameters

ADRange A/D input range in Volts, e.g., −5 5, or 0 10. Only certain values are
   supported, depending on the utilized board.
BoardNumber The A/D Board number as displayed by the InstaCal program.

11.1.4.4 States

None.

11.1.5 Data Translation Boards

The DTADC component supports analog/digital conversion boards by Data Translation.

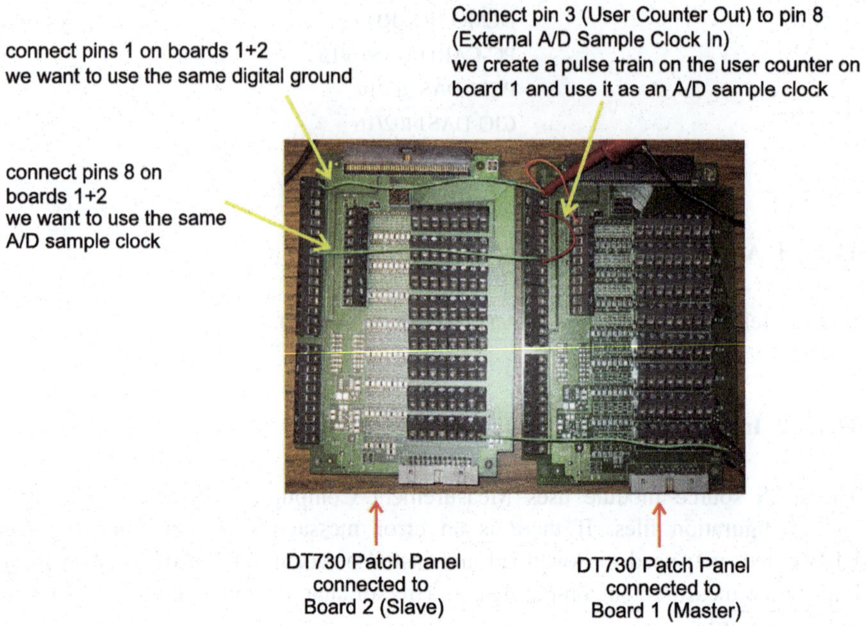

connect pins 1 on boards 1+2
we want to use the same digital ground

Connect pin 3 (User Counter Out) to pin 8
(External A/D Sample Clock In)
we create a pulse train on the user counter on
board 1 and use it as an A/D sample clock

connect pins 8 on
boards 1+2
we want to use the same
A/D sample clock

DT730 Patch Panel
connected to
Board 2 (Slave)

DT730 Patch Panel
connected to
Board 1 (Master)

Fig. 11.1 Photograph of DT730 patch panels with labeled connections

11.1.5.1 Authors

Gerwin Schalk & Dennis McFarland.

11.1.5.2 Installation

The DTADC supports multiple connected boards. To connect multiple boards, and
thus connect BCI2000 systems with up to 128 channels:

- Buy two DT3003 boards and two DT730 patch panels.
- Make the three required connections as shown in Fig. 11.1.
- In BCI2000, set **BoardName** to the name of board 1, **BoardName2** to the name
  of board 2, **SoftwareCh** to the total number of channels to be acquired, **Soft-
  wareChBoard1** to the number of channels to be acquired from board 1, and
  SoftwareChBoard2 to the number of channels to be acquired from board 2.
- (**SoftwareChBoard1** + **SoftwareChBoard2** has to equal **SoftwareCh**; **Soft-
  wareChBoard2** has to be smaller than **SoftwareChBoard1**).

11.1.5.3 Parameters

BoardName The name of the AD board.

BoardName2 The name of the second AD board, or 'none' if there is no second board present.

SoftwareChBoard1 The number of channels for board 1. Ignored in case of a single board.

SoftwareChBoard2 The number of channels for board 2. Ignored in case of a single board.

11.1.5.4 States

None.

11.1.6 Micromed

This module can be used to read data sent over TCP/IP from a Micromed acquisition unit running SystemPLUS Rev. 1.02.1056.

The BCI source module that receives the data from the acquisition is the server, so it must be in listening mode before the acquisition unit starts to save data (data is sent only during recording). Micromed sends 64 data packets per second for SD, 32 for LTM. The connection is reset if the computer running BCI2000 is too slow to read all packets. This happens gracefully in version 2.0. Make sure you have SystemPlus 1.02.1091 or higher. In this version, Micromed also gracefully closes the TCP/IP connection at BCI2000 errors.

A packet of data is structured in two sections. The first section is the header and the second is information data. The first packet sent from the client is the header of an EEG Micromed trace and is read for condition checking. If a note is added in SystemPLUS, it is sent as a note packet to the Source module. If a digital trigger is sent, the trigger code is sent to the Source module in two ways: the complete code and a bitmasked code. This makes it possible to send information on a condition and other information for offline analysis. The header packet and the data packets are in the same format as in a Micromed trace file. The note packet sends the complete note buffer, and is written in a textfile with the Micromed Sample Number.

11.1.6.1 Authors

Erik J. Aarnoutse, Rudolf Magnus Institute, UMC Dept. Psychiatry, Utrecht, The Netherlands, May 19, 2006.

11.1.6.2 Installation

To activate data transfer via TCP, it is necessary to add 3 registry keys on the acquisition unit under HKEY_CURRENT_USER/Software/VB and VBA Program Settings/Brain Quick - System 98/EEG_Settings:

1. `tcpSendAcq` is a string type key which must be set to `"1"` to activate data transfer via TCP, and to `"0"` to deactivate it.
2. `tcpServerName` is a string type key that represents the IP address of the computer to receive EEG data.
3. `tcpPortNumber` is a string type key that represents the port number used (e.g., `"5000"`).

11.1.6.3 Parameters

ConditionMask Bitmask. Digital trigger code AND Condition Mask equals conditions without extra information. When not used, set to 0xFF.
NotchFilter Power line notch filter.

- 0: disabled,
- 1: at 50 Hz,
- 2: at 60 Hz.

PacketRate Number of TCP/IP packets per second: 32 for LTM, 64 for SD.
Priority CPU Priority. Default = 1, set higher if the CPU load is too heavy.
SampleBlockSize The number of samples transmitted at a time in the BCI2000 system. If SampleBlockSize is a multiple of SamplingRate/64, data packets are merged. This way the BCI system can run on less than 64 sampleblocks per second, which saves CPU power.
SamplingRate The sampling rate set by the acquisition unit.
ServerAddress Address and port of the Micromed BCI Server. The port number can be set in the acquisition unit by changing the registry setting `tcpPortNumber`.
SignalType Numeric type of output signal:

- 0: int16,
- 3: int32.

Other values are not allowed. Use 0 for Micromeds 16 bit mode, 3 for 22 bit mode: only the lower 22 bits are used, but data packets are filled with 32 bit integers.
SourceCh The number of digitized and stored channels must match the number of channels in the acquisition unit.

11.1.6.4 States

None.

11.1.7 Modular EEG

This component supports devices built on the ModularEEG concept. The ModularEEG is a GPL-licensed EEG amplifier designed by Joerg Hansmann. The schematics, PCBs and design-documents were released to the OpenEEG – community in 2002. The ModularEEG is a low-cost EEG system that consists of a microcontroller-based digitizer board and one, two or three analog boards. Each analog board can capture two EEG signals. Thus, the Modular EEG can transmit two, four or six channels of EEG data. The transmission to the host PC or PDA is done via serial RS232 connection. USB converters can be used, and common baud rates are 56,700 or 115,200 bits per second. Be aware that the isolation-barrier for user safety is only 5 kV, which does not meet the criteria for a medical device. Thus, the ModularEEG may not be used for clinical applications. For details and filter specifications please refer to the design documentation (available at http://openeeg.sourceforge.net/doc/modeeg/modeeg_design.html).

The ModularEEG has a 10bit A/D converter and transmits EEG channel data in high byte/low byte format. The gain settings can be adjusted with potentiometers on the analog boards. A 14 Hz 250 μV calibration signal is provided by the digital unit. When calibration is set to ±250 μV, the value 0 corresponds to −250 μV and the value 1,024 corresponds to 250 μV. BCI2000 Signal Processing or any offline analysis routine can derive, as with any other BCI2000 Source module, sample values in μV by subtracting, from each stored sample, **SourceChOffset**, and multiplying it with **SourceChGain** for each channel.

11.1.7.1 Authors

- Christoph Veigl, FORTEC-Institute, Technical University Vienna, Austria.
- Gerwin Schalk, Brain–Computer Interface Research and Development Program, Wadsworth Center, New York State Department of Health.

11.1.7.2 Installation

11.1.7.3 Parameters

ComPort Number of the serial port the ModularEEG is connected to.
Protocol Transmission Protocol. There are currently three different protocols for data transmission from or to the EEG device.

- The oldest and most compatible protocol is called P2. It transmits all six channels (even in case there are only two connected) and uses unidirectional communication to the host computer. The P2 protocol is compatible to other EEG applications like "Electric Guru." A P2-packet consists of 17 bytes.
- P3 is a newer, more compact format. A six-channel data packet has 11 bytes, and the transmission of only four or two channels is possible. There are two firmware versions that are currently in experimental stage. The corresponding protocols are bi-directional and make it possible to send command frames to the ModularEEG.

SampleBlockSize Samples per digitized block.

SamplingRate The sampling rate for the EEG data. Currently, this is fixed to 256 Hz. New firmware versions will support adjustments of the sampling rate.

SimulateEEG When this option is selected, sine waves are generated instead of using real time EEG data. The amplitude and frequency of the sine waves can be adjusted by moving the mouse.

SourceCh The number of channels (2, 4, or 6).

11.1.7.4 States

None.

11.1.8 National Instruments

This Source module supports data acquisition from National Instruments analog/digital (A/D) boards. It was tested using National Instruments' 6.9.x drivers. It will not work with the newer MX drivers.

11.1.8.1 Authors

Gerwin Schalk, Wadsworth Center, NYSDOH.

11.1.8.2 Installation

11.1.8.3 Parameters

BoardNumber The NI-ADC board's device number.

11.1.8.4 States

None.

11.1.9 National Instruments MX

This NIDAQ_MX Source module supports data acquisition from National Instruments analog/digital (A/D) boards using MX driver version 8.5 or later. The driver also supports some traditional NIDAQ (legacy) boards; please refer to the official NI documentation for the complete list of supported boards (NI DAQMX 8.5 readme).

The BCI2000-compatible Source module (`NIDAQmx.exe`) can be used instead of any other Source module. It has been tested with DAQPad 6015 USB. In addition to standard parameters (i.e., **SampleBlockSize**, **SamplingRate**, **SourceCh**), this BETA driver is limited to acquiring 16 channels and that the terminal configuration is fixed to "Not Referenced Single Ended" on a bipolar range of ± 5 V. These settings may be changed at compilation time, and all the config instructions are placed in the `ADConfig()` function. Please refer to your hardware manual for supported settings.

11.1.9.1 Authors

Giulio Pasquariello, Gaetano Gargiulo, ©2008 DIET Biomedical unit, University of Naples "Federico II".

11.1.9.2 Installation

Include the `nidaqmx.lib` file in the project. This library file has already been converted from OMF to COFF using the omf2coff DOS utility contained in the Borland C Builder directory and is included in the ZIP file; the original NIDAQmx library is also included, and has been renamed to `NIDAQmx_orig.lib`.

11.1.9.3 Parameters

BoardNumber The NI-ADC board's device number.

11.1.9.4 States

None.

11.1.10 Neuroscan

The **NeuroscanADC** component implements the client side of the TCP/IP-based Neuroscan Acquire protocol. Thus, it may be used to interface BCI2000 with Neuroscan EEG systems.

NuAmps/SynAmps are EEG recording systems from Neuroscan, Inc., that are widely used in clinical settings. This section describes this support that consists of two components: A BCI2000-compatible Source module (`Neuroscan.exe`) and a command-line tool (*neurogetparams*). These components are described below and have been tested with Neuroscan Acquire version 4.3.1.

11.1.10.1 Using the NeuroscanADC Module

The BCI2000-compatible Source module Neuroscan.exe can be used instead of any other source module. In addition to standard parameters (i.e., **SampleBlock-Size**, **SamplingRate**, **SourceCh**), it only requests one Neuroscan-specific parameter (**ServerAddress**) that defines the IP address (or host name) and the port number of the Acquire server. An example for an appropriate definition is localhost:3999. Before starting BCI2000, you need to start *Acquire*, click on the *S* symbol in the top right corner (to enable the server) and start the server on any port. Please note that by default, *Acquire* suggests port 4000, which is a port used by BCI2000. You might use port 3999 instead. Once the server is running, you can start BCI2000. It is imperative that certain parameters (e.g., the number of channels) in BCI2000 match the settings in *Acquire*. You can read these settings from *Acquire* using the command line tool described in the following section. This command line tool can create a parameter file fragment that needs to be loaded on top of an existing parameter file every time a setting in *Acquire* is changed.

11.1.10.2 Authors

Gerwin Schalk, Wadsworth Center, New York State Department of Health, 2004.

11.1.10.3 Parameters

ServerAddress Address and port of the *Neuroscan Acquire* server, given in ad-dress:port format.

11.1.10.4 States

NeuroscanEvent1 An 8-bit state that encodes event information as sent over the *Neuroscan Acquire* protocol.

11.1.10.5 Neurogetparams Command Line Tool

This command line tool reads system settings from the *Neuroscan Acquire* server, displays them on a screen and creates a BCI2000 parameter file fragment if desired. It can be used as follows: neurogetparams -address localhost:3999 -paramfile test.prm. (The *Acquire* server has to be enabled before using this tool.) Once BCI2000 is configured correctly, this parameter file fragment needs to be loaded on top of the existing configuration to make sure that the settings match. You only need to repeat this procedure if you change settings in *Acquire* (e.g., such as the number of channels or the amplification). An example of the output display for this tool can be seen in Appendix B.

11.1.11 BrainAmp Systems

A Source module that interfaces BCI2000 with BrainAmp EEG systems (Brain-Products, Inc.) using the TCP/IP-based BrainAmp RDA interface.

11.1.11.1 Authors

Jürgen Mellinger (juergen.mellinger@uni-tuebingen.de), Thomas Schreiner (thomas.schreiner@tuebingen.mpg.de), Jeremy Hill.

11.1.11.2 Parameters

HostName IP address or name of the host to connect to, typically **localhost**.

SampleBlockSize The number of samples transmitted at a time. The *Vision-Recorder* appears to send one block every 40 ms, so **SampleBlockSize** should match **SamplingRate x 40 ms**.

SamplingRate The sampling rate as configured in the *VisionRecorder* program.

SourceCh The number of digitized and stored channels. This must match the setting in the *VisionRecorder* program, increased by 1 to account for an additional channel to hold marker information.

11.1.11.3 States

None.

11.1.11.4 getparams Tool

`getparams` is a command line utility that makes it possible to obtain appropriate Source module parameters from a host running BrainAmp's *VisionRecorder*.

On the target host, start the *VisionRecorder* program, check that RDA is enabled under *Configuration* → *Preferences*, and click the monitor (eye) button before running `getparams` with the host's IP address as the only parameter (when omitted, this defaults to **localhost**).

To direct the output into a file that can later be loaded into the Operator module's configuration dialog, append a redirection to the command:

```
getparams localhost > myparamfile.prm
```

11.1.12 Tucker-Davis Technologies

The **TDTclient** Source module interfaces BCI2000 with a Tucker-Davis Technologies (TDT) RX5 Pentusa system or RZ2 System. The TDT systems are capable of multi-channel (up to 64, or higher), high data transfer rates (more than 1 GB/s) allowing for high density EEG, ECoG, or even single-unit recordings. The system is highly configurable, allowing the user to use digital filtering and complex analysis on the hardware.

Currently, the TDTclient supports the following system configuration:

- TDT Pentusa/RX5 OR.
- TDT RZ2.
- GBIT or Optibit.
- ActiveX libraries installed (to install from source).

The obsolete Medusa system is not currently supported.

11.1.12.1 Authors

J. Adam Wilson (jadamwilson2@gmail.com), Department of Neurosurgery, University of Cincinnati.

11.1.12.2 Installation

The most current TDT drivers should be downloaded and installed from www.tdt.com. Also, make sure that the microcode on the hardware is updated to the latest version as well, or the TDT source module will not work.

Currently, the TDT ActiveX library must be installed on the local machine in order to use the TDTclient. It can be obtained from www.tdt.com, and it requires a password to install.

11.1.12.3 Parameters

CircuitPath The Circuit Path is the path to the TDT *.rco or *.rcx file to be used. Several files are included.

- chAcquire64.rco – Up to 64 channels from the RX5 system; requires a five processor system.
- chAcquire16.rco – Up to 16 channels from the RX5 system; requires a two processor system.
- chAcquire64_RZ2 – Up to 64 channels from the RZ2 system.

An eight-processor RZ2 system exists that should be capable of recording 256 channels. It is currently not supported.

DigitalGain If digital inputs are used on the front panel, they must be converted from a digital value (0 or 1) to a floating-point value with this scale factor.

FrontPanelList A list of front panel components to collect. The RZ2 system provides eight analog inputs and eight digital inputs which can be recorded as extra channels into BCI2000. To collect all eight analog channels, enter 1 2 3 4 5 6 7 8 for this value; to collect just digital channels, enter 9 10 11 12 13 14 15 16. Any combination of 1–16 is valid. Note: The number of values listed here plus **NumEEGchannels** *must* add to the value entered in the **SourceCh** parameter!

FrontPanelGain Analog inputs recorded from the front panel will have a different gain than those recorded on the EEG preamps, and need a different gain value to be recorded properly.

HPFfreq The corner frequency of the digital high-pass filter in the circuit. This can be set to 0 to not be used, but be aware that the Medusa pre-amps have an analog high-pass filter at 1.5 Hz built in.

LPFfreq The corner frequency of the digital low-pass filter. This can be set to any value, but it is recommended that the highest value used is at half of the sampling frequency (i.e., 256 for a sample rate of 512).

NumEEGchannels The number of EEG channels to acquire. These are channels recorded from the preamp. **NumEEGchannels** plus the number of entries in **FrontPanelList** must add to **SourceCh**.

notchBW This is the bandwidth of the 60 Hz notch filter. If set at 10, the bandstop filter will have corner frequencies of 55 and 65 Hz.

nProcessors The number of processors on the TDT system. This value must correspond to the rco file used in CircuitPath. For example, the chAcquire64_RZ2. rcx and chAcquire16.rco should have **nProcessors** set to 2, while chAcquire64.rco should use 5.

SampleRate The sampling rate of the system. Use the *TDTsampleRate* program to calculate the value to enter here. The TDT has a fixed rate of 24,414.0625 Hz, so the requested sample rate must be some integer fraction of this value. For example, a sample rate of 512 Hz is not possible, because it is impossible to find an integer that will divide 24,414.0625 to get 512. The closest rate to 512 Hz that can be used is 519.448, because 24,414.0625/519.448 = 47, which corresponds to the down-sampling factor.

TDTgain The amount of gain to add to the signals acquired from the amplifier. To convert to μV, this should be 1,000,000, unless the EEG pre-amplifier is used, in which case this should be 50,000, since the pre-amp adds a 20x gain.

11.1.12.4 States

None.

11.1.13 TMS Refa and Porti Systems

A Source module that interfaces BCI2000 to TMS Refa and Porti systems.

11.1.13.1 Authors

M.M. Span, ©RUG University of Groningen.

11.1.13.2 Parameters

The standard Source module parameters are used.

11.1.13.3 States

None.

11.1.14 BrainProducts V-Amp

The **vAmpADC** filter acquires data from up to four *V-Amp* or *FirstAmp* EEG amplifiers. The *V-Amp* is an amplifier/digitizer combination from Brain Products. Support for this device in BCI2000 consists of a BCI2000-compatible Source module (vAmpSource.exe).

11.1.14.1 Authors

J. Adam Wilson, University of Wisconsin–Madison & The Wadsworth Center, Albany, NY.

11.1.14.2 Hardware

The *V-Amp* consists of 16 independent 24-bit A/D converters that can sample at up to 2 kHz per channel (16 channels, plus two auxiliary) or 20 kHz per channel (four channels).

11.1.14.3 Installation

The necessary system drivers are located in the BCI2000/src/extlib/ brainproducts/vamp/ folder. To install the driver, follow these steps:

1. Plug in the *V-Amp* to a USB port. Windows should recognize it, and start to install drivers.
2. Depending on the version of Windows being used, a "Found New Hardware" dialog should appear. Do NOT search online for drivers; instead, select *Don't Search Online*.

3. When it asks you to insert the disc that came with the amplifier, select the option that allows you to browse for the driver.
4. Navigate to `BCI2000/src/extlib/brainproducts/vamp/`. It should find the driver files, and install the necessary software.
5. For the last step, copy the files `DiBpGmbH.dll` and `FirstAmp.dll` to the `BCI2000/prog/`folder.
6. You should now be able to run the `vAmpSource.exe` module in BCI2000.

11.1.14.4 Parameters

SamplingRate The sample rate of the system. All data are acquired either at 2,000 Hz or 20 kHz, and then decimated to the desired rate. Therefore, only rates that are integer divisors of the base rates are accepted. In 2,000 Hz mode, the valid rates are:

- 2,000
- 1,000
- 666.6
- 500
- 400
- 333.3
- 285.7
- 250
- 222.22
- 200

In 20 kHz mode, 10 times the above rates are valid. Prior to decimation, a second-order anti-aliasing Butterworth filter with a corner frequency of 0.45 times the sampling rate is applied to the signal. All sampling rates are supported for one or more amplifiers. If you are sampling at high rates and from multiple amplifiers, the CPU may be overloaded depending on the speed of your computer and the BCI2000 configuration. In case you are experiencing problems (e.g., data loss, jerky display, etc.), increase the SampleBlockSize so that you are updating the system less frequently (usually, updating the system 20–30 times per second is sufficient for most applications), and increase *Visualize* → *VisualizeSourceDecimation*. This parameter will decrease the number of samples per second that are actually drawn in the *Source* display.

DeviceIDs List of serial numbers (e.g., `70`) of all devices. Serial numbers of found devices will be listed in the BCI2000 log window. If you have more than one device, this list determines the order of the channels in the data file.

HPFilterCorner A high-pass digital filter for removing DC offsets. This occurs after data has been read into BCI2000, and before it is stored to disk.

SampleBlockSize Samples per channel per digitized block. Together with the sampling rate, this parameter determines how often per second data are collected, processed, and feedback is updated. For example, at 600 Hz sampling and a Sample-

BlockSize of 20, the system (e.g., source signal display, signal processing, and stimulus presentation) is updated 30 times per second.

SourceChDevices The number of channels acquired from each device. If there is only one device, this parameter has to equal **SourceCh**. For example, ' 16 8 ' will acquire channels from the first device listed under **DeviceIDs**, and eight channels from the second device listed under **DeviceIDs**. Data acquisition always starts at channel 1. The sum of all channels (e.g., 24 in this example) has to equal the value of **SourceCh**. In high-speed mode, this number may not be higher than 5 per device (four channels + digital).

SourceChList The list of channels that should be acquired from each device. The total number of channels listed should correspond to **SourceCh**. For more than one device, **SourceChDevices** determines how the **SourceChList** values are mapped to each device. For example if **SourceChDevices** = ' 8 8 ' and **SourceChList** = '1 2 3 4 13 14 15 16 5 6 7 8 9 10 11 12 ', then channels 1-4 and 13-16 will be acquired on the first device, and channels 5-12 will be acquired on the second device. These channels will be saved in the data file as 16 contiguous channels. The order of channels does not matter; i.e., '1 2 3 4 ' is the same as '2 3 1 4 '. The channels are always in ascending order on a single device. Channels may not be listed twice on a single device; e.g., entering '1 2 3 4 5 6 7 1 ' if **SourceChDevices** = '8 ' will result in an error. If this parameter is left blank (the default), then all channels are acquired on all devices. For the *V-Amp 16*, channels 1-16 are EEG channels, 17-18 are Auxiliary channels, and 19 contains the eight digital channels stored in bits.

AcquisitionMode

- If set to *analog signal acquisition*, the *V-Amp* records analog signal voltages (default).
- If set to *high-speed acquisition*, the *V-Amp* records analog signals at 20 kHz instead of 2,000 Hz.
- If set to *Calibration*, the signal output is a square wave test signal generated by the *V-amp* (which can be used to verify correct system calibration).
- If set to *Impedance*, impedance values are recorded instead of the signal.

This impedance test reports input impedances for each channel in a separate window. Values are color-coded according to the magnitude (Green for 0–5 kOhm, Orange for 5–30 kOhm, Red for 30–1000 kOhm, Purple for impedances >1 MOhm). Values are updated in real time. Channels are shown in columns, and devices in rows.

11.1.14.5 States

None.

11.1.14.6 Data Format and Storage

- The signals are originally recorded from the V-Amp as 32-bit integers, and converted into μV.
- The auxiliary input units are in Volts.
- **SourceChOffset** is required to be zero for all channels.

Any offline analysis routine can derive (as with any other BCI2000 Source module) sample values in μV by subtracting, from each stored sample, **SourceChOffset** (i.e., zero), and multiplying it with **SourceChGain** for each channel. If **SignalType** is set to `float32`, data samples are stored in units of μV. In this case, **SourceCh-Gain** should be a list of 1's (because the conversion factor between data samples into μV is 1.0 for each channel). Even when you specify values other than 0 and 1, BCI2000 will produce a consistent data file, i.e., values will be transformed before they are written to the file such that applying **SourceChOffset** and **SourceChGain** will reproduce the original values in μV.

11.2 Tools

11.2.1 EEGLabImport

11.2.1.1 Synopsis

EEGLabImport is a plugin for EEGLAB that allows importing data, including the recorded signal and event markers, that is stored in the BCI2000 data format. Data from multiple runs can be concatenated and imported simultaneously, greatly extending and simplifying the analysis of EEG- and ECoG-based data sets. Further information about EEGLAB can be found at: http://sccn.ucsd.edu/eeglab/.

11.2.1.2 Prerequisites

BCI2000import requires Matlab 7.0 or greater and a current version of EEGLAB. Depending on the size of the data sets, a large amount of RAM may also be required, due to resource requirements of Matlab.

11.2.1.3 Installing EEGLabImport

Installation of the plugin is straightforward, but may require compiling the `load_bcidat` MEX function for your platform.

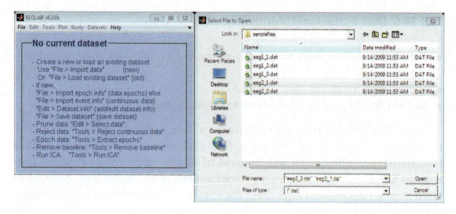

Fig. 11.2 The EEGLabImport program. Multiple files have been selected in this example

---

EEGLabImport

1. If you are using a 32-bit version of Windows, `load_bcidat.mexw32` is likely already compiled for your platform, and can be located in the `BCI2000/tools/mex/` folder. Copy this file to the `BCI2000/tools/EEGLabImport` folder, and skip to 5.
2. If you are using a platform for which a version of `load_bcidat` does not exist in the `/tools/mex` folder, the `load_bcidat` MEX file must be compiled.
3. Open Matlab, and navigate to `BCI2000/src/core/Tools/mex/`. Type `buildmex` at the command line; this will build the BCI2000 MEX functions for your platform.
4. Once the build process is complete, locate the `load_bcidat` file for your platform based on its extension (see the Matlab MEX documentation for more details), and copy it to `BCI2000/tools/EEGLabImport`.
5. Copy the folder `BCI2000/tools/EEGLabImport` to the EEGLab plugin folder, located at `eeglab/plugins`.

When EEGLAB is started, the BCI2000 data plugin will be automatically detected and installed.

11.2.1.4 Using EEGLabImport

Begin a Matlab session, and start EEGLAB by typing `eeglab` at the command prompt. EEGLabImport is detected and added to the menus automatically. Go to *File ➤ Import Data ➤ BCI2000 Data* to start the data import program. A file selection dialog is presented, in which multiple BCI2000 data files can be selected (see Fig. 11.2). The default file extension for BCI2000 data files is *.dat. Select one or more files from the desired folder, and press Open.

After pressing Open, the program loads the selected data, including the signal, states, and testing parameters such as the sampling rate. Next, a window will appear asking which BCI2000 states should be imported as EEGLAB events. Several important and common states will automatically be selected, but it is not required to import these. These events are used to generate signal epochs, which are then used for signal averaging and other analyses. Each event relates to a different experimental event.

Once the desired events are selected, EEGLabImport removes the undesired states, and finishes importing the data into EEGLAB. The final step is to give the dataset a name, and to optionally save the file in the EEGLAB format for future analysis.

11.2.1.5 Tutorial

This tutorial covers some of the basics of using EEGLAB to analyze data from a BCI experiment. The sample data sets used for this are included in the `BCI2000/data/samplefiles` folder. This data was collected during a P300 Speller training session, and EEGLAB will be used to compare the evoked response for the attend and ignore conditions.

Importing BCI2000 Data

1. Open Matlab, and navigate to the EEGLAB folder. Type `eeglab` to start EEGLAB.
2. In EEGLAB, go to *File* ➙ *Import Data* ➙ *From BCI2000 *.DAT*. Navigate to `BCI2000/data/samplefiles/`, and select `eeg2_1.dat` and `eeg2_2.dat`.
3. When EEGLAB asks which BCI events to import, select `StimulusCode`, `StimulusType`, and `StimulusBegin` (Fig. 11.3).
4. When asked, name the dataset `P300`.

Importing Channel Locations
In order to utilize many of EEGLAB's functions, it needs to know the location of each channel on the head. This is accomplished by loading a channel location file.

Importing Channel Locations

1. In EEGLAB go to *Edit* ➙ *Channel Locations*. This brings up the channel location edit window (Fig. 11.4).
2. Press the *Read Locations* button in the lower-left corner. Navigate to `BCI2000/data/samplefiles/`, and select `eeg64.loc`.
3. A window appears asking for the File Format; select `autodetect`, and press *OK*.
4. The channel locations are imported, and you can view the information for each electrode.
5. Press the *OK* button to close the channel edit window.

Fig. 11.3 In this example, the states `StimulusCode`, `StimulusType`, and `StimulusBegin` will be imported into EEGLAB

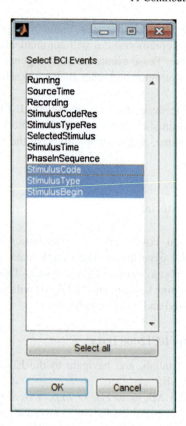

Importing Channel Locations (continued)

6. To verify that the channels are imported correctly, go to *Plot ➤ Channel Locations ➤ By name*. A plot showing each channel's location is displayed on a simple 2D view of a head.

Epoching Data

The next step is to epoch the data based on one or more events that were imported. The BCI task can be epoched in many ways, but the way it will be done here is by using the **StimulusBegin** state as the event marker. This BCI2000 state corresponds to the beginning of each flashing letter during the P300 Spelling task, and is therefore suitable to epoch the data for ERP analysis.

Importing Channel Locations

1. In EEGLAB, go to *Tools ➤ Extract Epochs* (Fig. 11.5).
2. The **Time-locking event** parameter determines which event should be used to create the epochs. Enter `StimulusBegin` in this field.

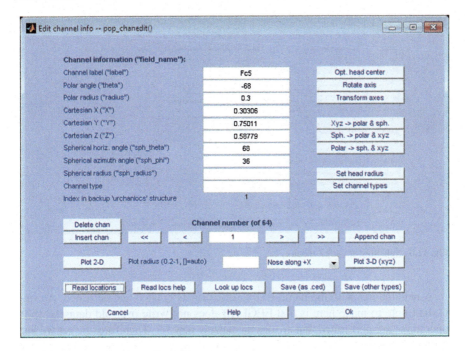

Fig. 11.4 Channel location edit window

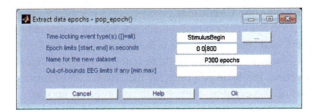

Fig. 11.5 Extracting data epochs

Importing Channel Locations (continued)

3. The **Epoch limits** parameter determines the time ranges of the data before and after each event. Enter 0 0.800 in this field, to keep the data starting with the flash onset to 800 ms after the flash.

4. In the **Name for the new dataset** field, enter P300 Epochs.

5. Press the *OK* button to finish.

6. A dialog asks to name the new dataset; enter P300 Epochs into the appropriate field.

7. The last dialog asks to remove the baseline from each epoch. Press *Ok* to remove the baseline from the entire epoch.

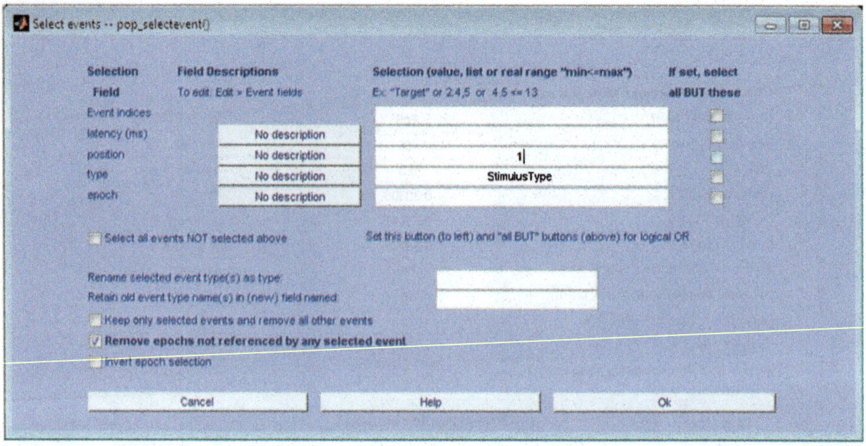

Fig. 11.6 Selecting attended epochs

A new dataset with the name P300 Epochs is created and activated. An important note to make here is that EEGLAB does not write over old datasets when operations are performed, but creates an entirely new set. Therefore, the set we just created is given an index of 2; the currently selected dataset can be changed in EEGLAB in the *Datasets* menu.

Extracting Task-Specific Conditions

During the P300 Spelling calibration test, the user is instructed to pay attention to specific letters while the letter matrix is flashing. The StimulusType state and event specifies whether the current flashing letter is the letter that the user should attend to. Therefore, the brain response for the **Attend** and **Ignore** conditions can be compared using this event by extracting the appropriate epochs.

Extracting the Attend Condition

1. In EEGLAB, select the *Edit → Select Epochs/Events* menu, which displays the Select Event dialog (Fig. 11.6).
2. We want to select the epochs in which the StimulusType equals 1. Therefore, enter StimulusType in the **type** field, and 1 in the **position** field.
3. The remaining options should not be changed. Press *Ok* to continue.
4. A *Confirmation* dialog is displayed warning that 160 epochs will be removed from the new data set. Press *Ok* to confirm.
5. A new dataset is created containing *only* the brain responses to the attended stimuli. A dialog asks what to name the dataset; enter P300 Attend into the appropriate field, and press *Ok* to continue.

Next, the brain signals for the Ignore condition will be extracted in a similar manner.

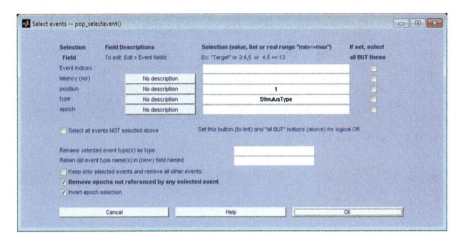

Fig. 11.7 Selecting ignored epochs

Extracting the Ignore Condition

1. First, the epoched dataset must be re-selected. Currently, dataset #3 is likely selected in EEGLAB, and this does not contain any "Ignore" data. Therefore, click *Datasets → Dataset 2: P300 Epochs* to select the correct dataset.
2. In EEGLAB, select the *Edit → Select Epochs/Events* menu, which displays the Select Event dialog (Fig. 11.7).
3. We want to select the epochs in which the StimulusType *does not* equal 1. Therefore, enter StimulusType in the **type** field, and 1 in the **position** field.
4. In the lower-left corner, click the box that says **Invert epoch selection**. This finds all of the epochs for which StimulusType equals 1, and selects *all others*, which corresponds to the Ignore condition. Press *Ok* to continue.
5. A *Confirmation* dialog is displayed warning that 40 epochs will be removed from the new data set. Press *Ok* to confirm.
6. A new dataset is created containing *only* the brain responses to the ignored stimuli. A dialog asks what to name the dataset; enter P300 Ignore into the appropriate field, and press *Ok* to continue.

Plotting ERP Maps

Next, we will plot the brain responses for the Attend condition.

Plotting the ERP Map

1. Select *Datasets → Dataset 3: P300 Attend*.
2. Select *Plot → Channel ERPs → With scalp maps*.
3. A dialog window appears with options for plotting the ERPs (Fig. 11.8).
4. For the **Plotting time range** field, keep the default value of 0 793.75.

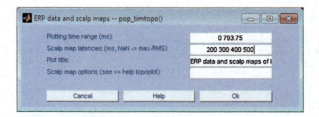

Fig. 11.8 Options for plotting ERPs

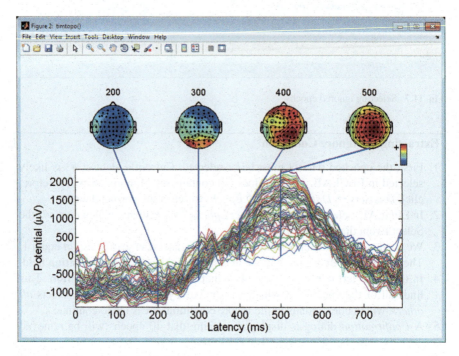

Fig. 11.9 ERP and scalp map for the Attend condition

Plotting the ERP Map (continued)

5. For the **Scalp map latencies** field, enter 200 300 400 500. These plot the scalp topographies at these latencies. Alternatively, leave this field blank, and EEGLAB will select the latency with the largest response.
6. Press *Ok*.
7. A plot of the averaged ERP for each channel is shown, along with the scalp maps at each latency (Fig. 11.9).

In Fig. 11.9, we can see that there is a large response centered over the Cz/Pz electrodes approximately 500 ms after the stimulus, as expected. It is also instructive to compare the Attend response to the Ignore response. To do so, select *Datasets* →

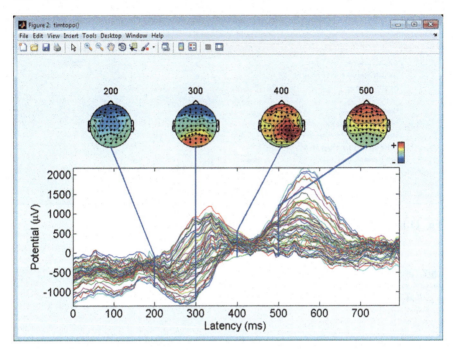

Fig. 11.10 ERP and scalp map for the Ignore condition

Dataset 4: P300 Ignore, and repeat the above steps to plot the ERP for the Ignore condition (Fig. 11.10).

Comparing ERP Responses

Finally, the ERPs for both conditions will be plotted to determine the locations and times of the largest differences.

Plotting the ERP Differences

1. Select *Plot → Sum/Compare ERPs*. The **ERP grand average** dialog appears (Fig. 11.11).
2. For **Datasets to average**, enter 3.
3. For **Datasets to average and subtract**, enter 4. This specifies that the average ERP for dataset 4 should be subtracted from the average ERP for dataset 3.
4. Make sure that the other options are not changed, and press *Ok*.

A plot of the ERPs at all locations are shown, including the attend (green) and ignore (blue) conditions, and the difference between them (purple) (Fig. 11.12).

To inspect an individual channel, click on its location in this plot. For example, click on the Cz axis to display an enlarged plot of the ERPs, as in Fig. 11.13.

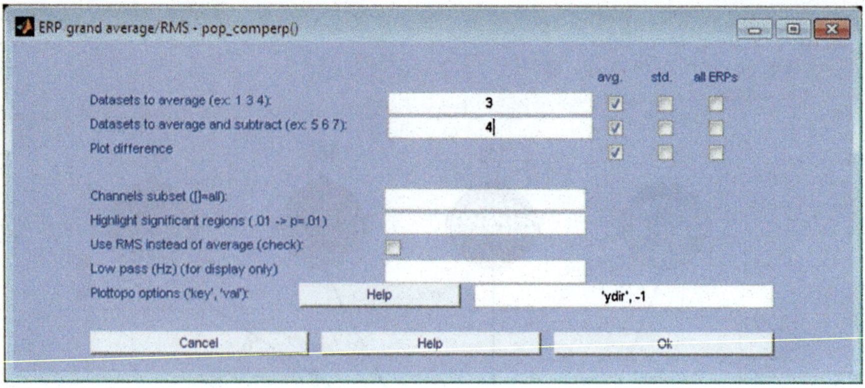

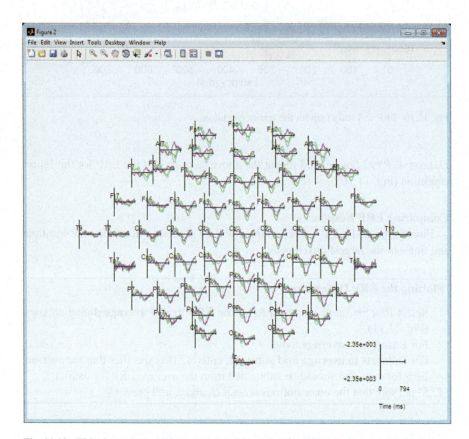

Fig. 11.11 The ERP grand average dialog

Fig. 11.12 ERPs for the Attend and Ignore conditions, and the difference between the ERPs

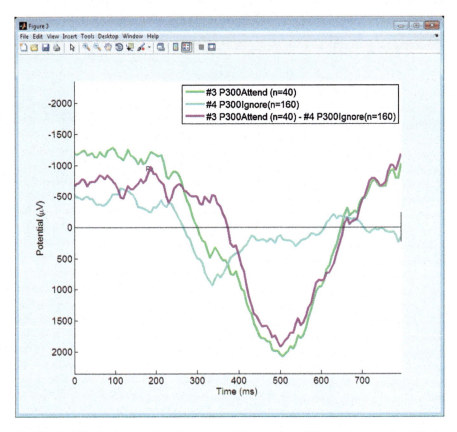

Fig. 11.13 ERPs for the Attend and Ignore conditions, and the difference between the ERPs on channel Cz

11.2.1.6 Further Information

This tutorial aimed at getting you started using EEGLAB for analyzing BCI2000 data. Detailed information on using the EEGLAB software is provided by the EEGLAB tutorial, which is available at the EEGLAB homepage at http://www.sccn.ucsd.edu/eeglab/.

Appendix A
The `USBampGetInfo` Command Line Tool

The USBampGetInfo tool prints information about the filter capabilities of attached g.USBamp amplifiers. This information can be used to configure the **USBamp-Source.exe** module with proper filter settings.

```
*********************************************
BCI2000 Information Tool for g.USBamp
*********************************************
(C)2004 Gerwin Schalk
     Wadsworth Center
     New York State Department of Health
     Albany, NY, USA
*********************************************
Amp found at USB address 1 (S/N: UA-200X.XX.XX)
Printing info for first amp (USB address 1)

Available bandpass filters
===================================
num| hpfr  | lpfreq | sfr | or | type
===================================
000|  0.10 |   0.0  |  32 |  8 | 1
001|  1.00 |   0.0  |  32 |  8 | 1
002|  2.00 |   0.0  |  32 |  8 | 1
003|  5.00 |   0.0  |  32 |  8 | 1
004|  0.00 |  15.0  |  32 |  8 | 1
005|  0.01 |  15.0  |  32 |  8 | 1
006|  0.10 |  15.0  |  32 |  8 | 1
007|  0.50 |  15.0  |  32 |  8 | 1
008|  2.00 |  15.0  |  32 |  8 | 1
009|  0.10 |   0.0  |  64 |  8 | 1
010|  1.00 |   0.0  |  64 |  8 | 1
011|  2.00 |   0.0  |  64 |  8 | 1
012|  5.00 |   0.0  |  64 |  8 | 1
013|  0.00 |  30.0  |  64 |  8 | 1
014|  0.01 |  30.0  |  64 |  8 | 1
```

```
015|   0.10  |   30.0  |    64  |  8  |  1
016|   0.50  |   30.0  |    64  |  8  |  1
017|   2.00  |   30.0  |    64  |  8  |  1
018|   0.10  |    0.0  |   128  |  8  |  1
019|   1.00  |    0.0  |   128  |  8  |  1
020|   2.00  |    0.0  |   128  |  8  |  1
021|   5.00  |    0.0  |   128  |  8  |  1
022|   0.00  |   30.0  |   128  |  8  |  1
023|   0.00  |   60.0  |   128  |  8  |  1
024|   0.01  |   30.0  |   128  |  8  |  1
025|   0.01  |   60.0  |   128  |  8  |  1
026|   0.10  |   30.0  |   128  |  8  |  1
027|   0.10  |   60.0  |   128  |  8  |  1
028|   0.50  |   30.0  |   128  |  8  |  1
029|   0.50  |   60.0  |   128  |  8  |  1
030|   2.00  |   30.0  |   128  |  8  |  1
031|   2.00  |   60.0  |   128  |  8  |  1
032|   0.10  |    0.0  |   256  |  8  |  1
033|   1.00  |    0.0  |   256  |  8  |  1
034|   2.00  |    0.0  |   256  |  8  |  1
035|   5.00  |    0.0  |   256  |  8  |  1
036|   0.00  |   30.0  |   256  |  8  |  1
037|   0.00  |   60.0  |   256  |  8  |  1
038|   0.00  |  100.0  |   256  |  8  |  1
039|   0.01  |   30.0  |   256  |  6  |  1
040|   0.01  |   60.0  |   256  |  8  |  1
041|   0.01  |  100.0  |   256  |  8  |  1
042|   0.10  |   30.0  |   256  |  8  |  1
043|   0.10  |   60.0  |   256  |  8  |  1
044|   0.10  |  100.0  |   256  |  8  |  1
045|   0.50  |   30.0  |   256  |  8  |  1
046|   0.50  |   60.0  |   256  |  8  |  1
047|   0.50  |  100.0  |   256  |  8  |  1
048|   2.00  |   30.0  |   256  |  8  |  1
049|   2.00  |   60.0  |   256  |  8  |  1
050|   2.00  |  100.0  |   256  |  8  |  1
051|   5.00  |   30.0  |   256  |  8  |  1
052|   5.00  |   60.0  |   256  |  8  |  1
053|   5.00  |  100.0  |   256  |  8  |  1
... continued for sampling rates up to 4800 Hz


Available notch filters
=====================================
num| hpfr  | lpfreq |  sfr  | or | type
=====================================
000| 48.00  |   52.0  |  128  |  4  |  1
001| 58.00  |   62.0  |  128  |  4  |  1
002| 48.00  |   52.0  |  256  |  4  |  1
003| 58.00  |   62.0  |  256  |  4  |  1
```

```
004|  48.00  |    52.0  |   512  |   4  |  1
005|  58.00  |    62.0  |   512  |   4  |  1
006|  48.00  |    52.0  |   600  |   4  |  1
007|  58.00  |    62.0  |   600  |   4  |  1
008|  48.00  |    52.0  |  1200  |   4  |  1
009|  58.00  |    62.0  |  1200  |   4  |  1
010|  48.00  |    52.0  |  2400  |   4  |  1
011|  58.00  |    62.0  |  2400  |   4  |  1
```

Appendix B
The **neurogetparams** Command Line Tool

```
BCI2000 Parameter Tool for Neuroscan Acquire V4.3
********************************************************************
(C)2004 Gerwin Schalk and Juergen Mellinger
        Wadsworth Center, New York State Dept of Health, Albany, NY
        Eberhard-Karls University of Tuebingen, Germany
********************************************************************
Signal Channels: 32
Event Channels:  1
Block Size:      40
Sampling Rate:   1000
Bits/Sample:     16
Resolution:      0.168 muV/LSB
Parameter file test.prm successfully written
```

G. Schalk, J. Mellinger, *A Practical Guide to Brain–Computer Interfacing with BCI2000*,   253
© Springer-Verlag London Limited 2010

Index

Lightning Source UK Ltd.
Milton Keynes UK
UKOW06f0608120517
301016UK00009B/289/P